SpringerBriefs in Optimization

SpringerBriefs in Optimization showcases algorithmic and theoretical techniques, case studies, and applications within the broad-based field of optimization. Manuscripts related to the ever-growing applications of optimization in applied mathematics, engineering, medicine, economics, and other applied sciences are encouraged.

More information about this series at http://www.springer.com/series/8918

Athanasia Karakitsiou

Modeling Discrete Competitive Facility Location

Springer

Athanasia Karakitsiou
Industrial Logistics
ETS Institute
Luleå University of Technology
Luleå, Sweden

ISSN 2190-8354 ISSN 2191-575X (electronic)
SpringerBriefs in Optimization
ISBN 978-3-319-21340-8 ISBN 978-3-319-21341-5 (eBook)
DOI 10.1007/978-3-319-21341-5

Library of Congress Control Number: 2015945651

Mathematics Subject Classification (2010): 90-XX, 90BXX, 90CXX, 91XX

Springer Cham Heidelberg New York Dordrecht London

Printed on acid-free paper

Springer International Publishing AG Switzerland is part of Springer Science+Business Media (www.springer.com)

This work is dedicated to my mother
Chrysoula

Preface

Competitive facility location problems are concerned with the following situation: a firm wants to locate a predefined number of facilities to serve customers located in some region where there already exist (or will be) other firms offering the same service. Both new and existing firms compete for optimizing their market share of profit. A discrete version of such problems arises when it is assumed that there are a finite number of candidate locations and the markets consist of point demands.

This monograph surveys and provides insights regarding modeling complexity and algorithmic approaches to discrete competitive location problems. Not only competition on equal term but even competition between market leader and followers is considered in this study. Thus, bi-level optimization methodology is emphasized and further developed. In traditional location modeling, assignment of customer demands to supply sources is made for which the associated costs target the firm and not the customers. However, in many real-world situations, the cost incurs the customers. Moreover, there may be customer competition for the provided services. Thus, a new methodological framework is needed in order to encompass such considerations into the modeling and solution process. The monograph offers initial directions for further research and development along these lines.

This monograph is intended for graduate students and researchers in the field of mathematics, computer science, operational research, and game theory working with location decisions from the point of view of mathematical programming and optimization.

I would like to thank prof. Panos M. Pardalos for inviting me to contribute to this series, Springer and especially Razia Amzad for their support in making the publication of this volume possible. I would also like to thank the ETS Institute, Luleå University of Technology for providing me with the necessary means and resources to complete this work.

Luleå, Sweden Athanasia Karakitsiou

Contents

Chapter 1
Introduction

Facility location analysis is one of the most well-studied areas of the operations research [19, 36, 37, 77]. In the basic model, there is a predefined cost for opening a facility and also connecting a customer to a facility, the goal is to minimize the total cost. The typical facility location problem assumes that the locating facility is either a price taker or a monopolist, so that the market competition is neglected among the companies. However this simplified assumption does not fit in most real-life situation and the need arises to incorporate competition among the decision-makers. Indeed, competitive location models additionally incorporate the fact that location decisions have been or will be made by independent decision-makers who will subsequently compete with each other for market share, profit maximization, etc. [18]. In addition, the assignment of customers being served by these facilities and how these facilities are connected with each other are interesting decisions considered within the problem.

It is widely accepted that the competitive location analysis was initiated by Hotelling [33]. In his two ice cream vendors game he examined location policies of interdependently acting duopoly in a linear market of a given length. The distribution of buying power along the line segment is assumed uniform. Each customer has an inelastic demand for the good and pays the transportation cost of obtaining the good. Therefore, he patronizes the nearest facility in order to minimize his expenditures. He proved that a "back to back" location in the middle of the market constitutes a long run equilibrium. Since then a vast number of publications have been devoted to the subject. Thus, different classification efforts with respect to multiple components have been proposed in the literature, see, for example, [17, 20, 43, 70] among others. However, it is generally accepted that any effort of categorization of a competitive facility location (CFL) should be based on the following elements.

A. Karakitsiou, *Modeling Discrete Competitive Facility Location*,
SpringerBriefs in Optimization, DOI 10.1007/978-3-319-21341-5_1

1.1 Spatial Representation

The type of space where a problem is explicitly defined determines the set of methods that can be applied. It is possible to distinguish three different location spaces: *Continuous*—where the potential location of the facilities can be anywhere in the plane [21, 59]. *Discrete*—where facilities are allowed to locate at a finite set of possible locations and *network*—where any point on a network is suitable for location [70].

Once the location space has been defined, a measure of distance over the search space is required. In continuous problems, some distance functions derived, for example, from the Manhattan distance or the Euclidean distance are employed. In contrast, distances in network problems are measured using, typically, the notion of the shortest path on the network. For discrete problems, a matrix of inter-distances is usually given.

From the optimization point of view, the techniques used to cope with these problems also differ. Continuous location problems are, for most of the cases, nonlinear optimization problems, while discrete and network location problems are integer programming/combinatorial optimization problems.

1.2 Nature of Competition

This category includes the following classes of CFL. (a) *Static CFL problems,* [57] this class of problems assumes that a firm enters in a market, where operate some existing firms, aiming at choosing the location of p facilities such as to attract the maximum market share. The new competitor enters in the market by having full and in advance information about the characteristics and the strategies of the existing firm(s). It is assumed further that this information is fixed and no reaction is expected from the existing competitor(s). When the assumption of the non-reacting competitor(s) is eliminated, that is, when the competitor(s) react(s) after the new firm's entrance new classes of CFL arise: (b) *dynamic* and (c) *sequential* location problems [14, 28]. The competing firms make a location decision simultaneously in the first case, whereas there is a hierarchy in the decision-making process in the second case. The sequential location completion is typically formulated as a case of Stackelberg game characterized by two types of players, *leaders* who react first and make location decision taking into account the reaction of their competitors and *followers*, who make their decisions based on the previous decisions of the leaders. The solution procedure employed usually is this of finding the *Stackelberg equilibrium.* On the other hand, in simultaneous location games, the Nash equilibrium, that is, the location choices where no player has the incentive to unilaterally change location constitute the solution of the problem.

1.3 Number of New Facilities

It is customary to differentiate between single-facility location problems and multifacility problems. In the former, only one facility is to be located ($p = 1$), while in the latter, several facilities are to be sited ($p > 1$). The parameter p may be exogenous or may be determined by the optimal solution of the problem. When locating multiple new facilities, interactions between them may have to be considered, usually through the interdistances between the new facilities. Furthermore, in case $p > 1$, in some problems one must also decide how customers are allocated to facilities, leading to location-allocation problems. In some cases, facility planners decide which facility will serve customers, as it happens in public sectors (e.g., when school boards allocate students to schools), while in other cases, the clients are the ones who decide where to shop or obtain a service (e.g., restaurants, supermarkets, gas stations, etc.).

1.4 Features of Customers

In the CFL problems customers can be originated in a finite set of point in the plane or at vertices in the network or be dispersed over a region according to some distribution.

Another important characteristic is the behavior of the customers and consequently their demand. The demand can be either *inelastic*, that is, known with certainty and fixed or *elastic*, that is, it may vary depending on prices, traveling expenses, quality level, etc.

The patronizing behavior of the customers is also a very important factor when it comes to classification. The main question here is if the customers are free to choose the facility to be served. If this is the case, then his patronizing behavior may be *deterministic* if he satisfies his demand by a single facility to which is attracted more. If the customer allocates his demand to any existing facilities based on their competitive characteristics, then his patronizing behavior is characterized as *uncertain* or *probabilistic*. In order to measure the attraction of the customer toward a specific facility, the so-called *attraction function* is employed. This function measures the effect of the characteristic of the competitors, such as distance, price, quality of service provided to the customer' choice.

1.5 Objectives

Traditionally, the CFL model objectives have fallen into two categories: *pull* and *push* objectives. In the former case, it is considered that the closer the facilities are to the customers, the better the value of the objective function, while in the latter case, facilities are undesirable, so customers want to maintain them as far away as possible. One main pull objective is the maximization of market share, these problems are usually named *capture problems*. In this case, customers shop at the cheapest facility and take care of carriage. Another important pull objective is the minimization of the transportation cost. The firm decides to pursue a policy of delivered pricing, i.e. it will pay for the transportation costs. These type of problems are referred to as *median* in a network setting or *Weber* in planar-continuous models. Another possible objective is the maximization of the profit.

1.6 Price Setting

Another important component of a CFL is the incorporation of the pricing decision. Traditionally, the pricing policies which are utilized are either a *mill price policy* (the seller sets a factory price, the same for all the customers in the market, and the buyer takes care of carriage) or a *delivered price* policy (the seller charges a specific price in each market area, which includes the freight cost, and takes care of transport). Price can be set either simultaneously or sequentially.

Chapter 2
Static Competitive Facility Location Problems

In this section we review some of the most important contribution to the field of static competitive facility location problem. The main assumptions of the models presented in this section can be summarized as follows:

- the existing competition is known and fixed,
- the product sold is homogeneous,
- the customers' decision in patronizing the facility is based entirely on the distance traveled.
- the unit costs are the same in all stores regardless of ownership

2.1 Static Deterministic Competitive Facility Location Problems

The *maximum capture problem* (MAXCAP), first introduced in [60] and extensively reviewed in [70]. The MAXCAP attempts to define the location of a fixed number of facilities (p servers) belonging to a firm (Firm A) in a network where there already exist q other facilities competing for customers. It is assumed that the q existing facilities belong to only one competing firm (firm B). A node of the network represents a local market with fixed demand. Competition is based on distance. A marker captured by the facility locates closer to it. If two facilities are located in equal distance from a market, then their capture is divided into equal parts. The objective of the entering firm is to maximize its market capture. The formulation of the MAXCAP is as follows [60]:

$$\max \sum_{i \in I} w_i y_i + \sum_{i \in I} \frac{w_i}{2} z_i \tag{2.1}$$

A. Karakitsiou, *Modeling Discrete Competitive Facility Location*,
SpringerBriefs in Optimization, DOI 10.1007/978-3-319-21341-5_2

$$st \quad y_i \leq \sum_{j \in P_i} x_j, \quad \forall i \in I \tag{2.2}$$

$$z_i \leq \sum_{j \in T_i} x_j, \forall i \in I \tag{2.3}$$

$$y_i + z_i \leq 1, \forall i \in I \tag{2.4}$$

$$\sum_{j \in J} x_j = p, \tag{2.5}$$

$$y_i \in \{0, 1\}, \forall i \in I \tag{2.6}$$

$$z_i \in \{0, 1\}, \forall i \in I \tag{2.7}$$

$$x_j \in \{0, 1\}, \forall j \in J \tag{2.8}$$

where the variables are defined as follows:

$$y_i = \begin{cases} 1, \text{ if firm A captures demand node } i \\ 0, \text{ otherwise} \end{cases}$$

$$x_j = \begin{cases} 1, \text{ if A locates a facility at node } j \\ 0, \text{ otherwise} \end{cases}$$

$$z_i = \begin{cases} 1, \text{ if demand node } i \text{ is divided between A and B} \\ 0, \text{ otherwise} \end{cases}$$

and the parameters are:

- I the set of the demand areas, $i \in I$
- J the set of potential location sites, $j \in J$
- w_i the demand at node i
- P_i denotes the set of sites that firm A would patronize if she opens a new facility there.
- T_i denotes the set of sites of firm A that are located closed with the currently patronized competitor's facility.

The objective function (2.1) maximizes the market capture of the Firm A. Constraints (2.2) quarantine that if one of the variables $x_j = 1$ then the capture variable is allowed to be 1. The set of constraints (2.3) have similar meaning for the case where firm A locates its facility close to competitor's site. Constraints (2.4) ensure a demand note i can have only three stages, it can be either captured fully by firm A ($y_i = 1$ and thus $z_i = 0$) or be divided between both firms ($z_i = 1$ and thus $y_i = 0$), or be captured by firm B ($y_i = 0$ and $z_i = 0$), while constraints (2.5) defines the number of facilities located by firm A. Finally constraints (2.6)–(2.8) force the decision variables to take only two values.

In order to solve the problem ReVelle [60] used linear programming and branch and bound in a very small network. In [70] the algorithm of market capture (AMACA), a one-opt heuristic, is employed to solve large problems. This algorithm can be stated as follows [70]: Given an initial location of the p Firm's servers and knowing the location of the competitors, the capture that Firm achieves in the market can be obtained. Then, from a list of Firm A's servers, the first member of the list is picked and its location is moved to an empty node. The new capture can be computed and compared to the market capture achieved before the one-opt trade. If the objective has improved, the new set of locations is kept as the current solution. On the other hand, if the capture obtained before the trade is higher, this exchange is not considered. These one-opt trades are executed for all pairs of empty nodes and facilities. If at the end of all trades the objective has improved, then the procedure is repeated. The heuristic procedure ends whenever there is no improvement in the objective on completion of all exchanges possible. The AMACA heuristic procedure is very efficient in terms of computing time but does not guarantee optimality.

The MAXCAP has formed a foundation of a series of models [24]. The initial formulation of the MAXCAP has been through several modifications by relaxing some of its basic assumption. In [61] an extension of MAXCAP to allow both reallocation of existing and location of new facilities of Firm A is considered. The accompanying problem formulation, also known as *maximum capture problem with reallocation* (MAXRELOG), can be derived easily by replacing constraints (2.5) with the following:

$$\sum_{j \in J} x_i = p + s \tag{2.9}$$

$$\sum_{j \in J_A} x_j = p - r \tag{2.10}$$

where J_A is the set of the existing p facilities of the firm A, r, $(r \leq p)$ the facilities to be reallocated, and s the facilities to be located.

In the modification presented by Serra and ReVelle [71] not only location but also pricing decision is considered. The PMAXCAP assumes that in a spatial market there is an existing firm (firm B) which operates with q facilities. A new firm (firm A) wants to enter into the market by locating n facilities and competes with firm B on the basis of price and location. Both firms follow a uniform milling pricing policy, that is, all facilities of a given firm charge the same price (p^A and p^B, respectively) and customers incur the transportation cost. Thus, customers' decision on which firm to visit are based on transportation cost and price. If d_{ij} is the distance between market i and a facility j and t is the unit transportation cost, then the price π_i faced by customers in market i is $\pi_i = p + td_{ib_i}$, where b_i is the closed facility to i, and the demand function of a market i shopping at is closest facility is $D_i = p + td_{ib_i}$. Therefore, the demand function of firm A for each market i is defined as:

$$D_i^A(\pi^A, \pi^B)) = \begin{cases} D_i(\pi^A), \text{ if } p^A + td_{ib_i^A} < p^B + td_{ib_i^B} \\ 0, \qquad\quad \text{ if } p^A + td_{ib_i^A} \geq p^B + td_{ib_i^B} \end{cases} \tag{2.11}$$

If at each facility entails a production cost of the form $f_j + vq$, where f_j is the fixed setup cost which varies with the location and v and q are the unit production cost and the level of production at the site, respectively, then the profit for firm A is given by

$$\Pi^A = (p^A - v)\sum_{i\in I} D_i^A(\pi^A, \pi^B) - \sum_{j\in J^A} f_j \tag{2.12}$$

Consequently, the problem of firm A is to determine its location set J^A and the price p^A so as to maximize this profit.

The demand function can be either elastic or inelastic with respect to prices. If the demand is completely inelastic to price changes, then $D_i = a_i$, where a_i is the total quantity purchased by the market i. Firm A has to solve the following problem [71]:

$$\max\ (p^A - v)\left(\sum_{i\in I} a_i y_i^A\right) - \sum_{j\in J} f_j x_j^A \tag{2.13}$$

$$st\ \ y_i^A \leq \sum_{j\in P_i} x_j^A, \forall i \in I, \tag{2.14}$$

$$\sum_{j=1}^{n} x_j^A = n^A, \tag{2.15}$$

$$y_i^A \in \{0, 1\}, \forall i \in I, \tag{2.16}$$

$$x_j^A \in \{0, 1\}, \forall j \in J \tag{2.17}$$

where n^A is the number of firm's A located facilities and the variables are defined as in the MAXCAP problem. If the demand is elastic to price changes, then by imposing variables

$$x_{ij} = \begin{cases} 1, \text{if demand area } i \text{ is assigned to a facility } j \text{ belonging to A} \\ 0, \text{otherwise} \end{cases}$$

$$z_j = \begin{cases} 1, \text{if firm A locates a facility at node } j \\ 0, \text{otherwise} \end{cases}$$

the problem can be formulated as

$$\max\ (p^A - v)\sum_{i\in I}\sum_{j\in J} D_i^A(\pi^A, \pi^B)x_{ij} - \sum_{j\in J} f_j z_j \tag{2.18}$$

$$st\ \sum_{j\in J} x_{ij} = 1, \forall i \in I, \tag{2.19}$$

$$x_{ij} \le z_j\ \ \forall i \in I, \forall j \in J, \tag{2.20}$$

$$\sum_{j\in J} z_j = n^A, \tag{2.21}$$

$$x_{ij} \in \{0, 1\}, \forall i \in I, \tag{2.22}$$

$$z_j \in \{0, 1\}, \forall j \in J \tag{2.23}$$

A hybrid heuristic to solve both the problems is presented in [71]. This heuristic has a construction phase and two improvement phases.

2.2 Static Probabilistic Competitive Facility Location Problems

Models presented in the previous sections assume that the distance traveled is the only criterion affecting the patronizing behavior of the customers. Models presented in the previous sections assume that the distance traveled is the only criterion affecting the patronizing behavior of the customers. However, in more realistic situations, customers consider other attributes of the facilities during their decision-making process such as size, quality of product, and service provided.

Huff [34] suggested to measure the attraction felt by customers for a facility as a measure of his patronizing probability. In his model the attraction felt by a customer at zone i towards a facility j located at place x_j is proportional to the size of the facility and inversely proportional to a power of the distance between zone i and x_j. A general formulation of the attraction function is given by

$$u_{ij} = \frac{A_j}{f(d_{ij})} \tag{2.24}$$

where A_j is the attractiveness or quality of the facility j and f is a non-decreasing function of distance.

In the multiplicative competitive interaction (MCI) model of Nakanishi and Cooper [54] different attributes of the facility were used together by taking their product after weighting them by raising each to a power

$$u_{ij} = \prod_{k=1}^{s} x_{ijk}^{\beta_k} \tag{2.25}$$

where s is the set of facility's attributes, x_{ijk} is the kth attribute describing a facility j by customers at i, and β_k is the weight of the kth attribute.

The additive utility function is utilized in [16]. A general form of this function can be

$$U = \sum_{k=1}^{s} \beta_k f_k(x_k) \tag{2.26}$$

where x_k is the kth attribute and β_k its associated weight.

Other models (see, for example, [32]) make use of the exponential attraction function which is generally given by

$$A_{ij} = a_j^{\alpha} e^{-\beta d_{ij}} \tag{2.27}$$

where a_j measures the quality of the facility j and α, β are parameters determined empirically.

Berman and Krass [9] and Aboolian et al. [1] develop a spatial interaction model with variable expenditure where demand *canibalization* and *market expansion* are taken into account. Canibalization occurs when new facilities capture some of the demand from pre-existing facilities, while market expansion occurs when the total available consumer demand in the area increases as a result of opening new facilities. The attraction function they employed is a Huff based rule.

In [9] it is assumed that customer demand is concentrated at $N = 1, \ldots, n$ discrete points with point i containing w_i customers and d_{ij} is the distance from point i to point j. $X \subset N$ is the set of potential facility locations. Each location can support at most one facility. $E^0 \subset X$ is the set of existing facilities and $\bar{E} = X - E^0$ is the set of locations available for new facilities. The goal is to define the optimal set of locations of m new facilities, $E^1 \subset \bar{E}$, $|E| = m$. It is assumed that $w_j = 0, \forall j \in X$. If u_{ij} is the utility for customer at point $i \in N$ of a facility at node $j \in X$, then vector $\mathbf{u}_i$ (with $u_{ij} = 0$ if $j \notin E^0 \cup E^1$) summarizes the utility gained by a customer at i from all the facilities and the quantity $U_i(E^1) = \sum_{j \in E^0 \cup E^1} u_{ij}$ for $i \in N$ represents the total utility derived by a customer at i from the service provided by all facilities in X. Finally, the expenditure function $g_i(U_i(E^1))$ measures the total expenditures by customer at i. It is assumed to be non-negative and non-decreasing with respect to $U_i(E^1)$.

Depending on the form of the expenditure function the competitive facility location problem can take several forms. If it is a bounded linear function of the form

$$g_i(U_i(E^1)) = \begin{cases} a_i U_i(E^1)) & \text{if } U_i(E^1)) \leq U_i' \\ a_i U' & \text{if } U_i(E^1) > U' \end{cases} \quad i \in N, \tag{2.28}$$

where a_i, U_i' are parameters with $0 < a_i \leq 1/\hat{U}_i$ and $0 < U_i' \leq \hat{U}_i$. $\hat{U}_i$ is the maximum utility that can possibly be derived by a customer at i, then, by introducing binary variable $x_j, j \in X - E^0$ indicating whether the facility is located at j ($x_j = 1$) or not ($x_j = 0$), the following competitive problem should be considered:

$$\max z(E^1) = \sum_{i \in N} w_i g_i \sum_{j \in X - E^0} \frac{u_{ij} x_j}{\sum_{k \in X - E^0} u_{ik} x_k + \sum_{k \in E^0} u_{ik}} \tag{2.29}$$

$s.t$

$$g_i \leq a_i U_i', i \in N. \tag{2.30}$$

$$g_i \leq a_i \sum j \in X - E^0 u_{ij} x_j + \sum_{j \in E^0} u_{ij}, i \in N, \tag{2.31}$$

$$\sum_{j \in x - E^0} x_j \leq m, \tag{2.32}$$

$$g_i \leq 0, i \in N, x_j \in \{0, 1\}, j \in \bar{E} \tag{2.33}$$

A greedy heuristic is proposed for the solution of the problem where instead of recomputing the objective function anew at every iteration.

In [1] a special case of the model in [9] is considered. It is assumed that customers' demand is a function of the total utility derived by each customer from the service offered by the facilities, satisfied certain concavity assumption. The concavity of the demand allows them to formulate the problem as a nonlinear Knapsack problem with a particular form of the objective function. Due to complexity of the objective function it is hard to solve large instances of the problem. In [1] the TLA (Tangent Line Approximation) algorithm is developed for the solution of the problem. TLA is based on over-approximating the nonlinear objective function with piecewise linear function.

Another part of the literature is devoted to the determination of the optimal values of the facilities' attractiveness. The model presented by Aboolian et al. [2] seeks to simultaneously optimize location and design decisions for a set of new facilities. The facilities compete for capturing customers' demand with each other and with existing facilities. The demand of the customers at $q(U)$ is assumed to be a concave function increasing with respect to the utility of the service provided by the facilities, that is, demand increases as the utility of the service provided by the facility increases. Specifically it is assumed that a non-decreasing function $q(U_i)$ is specified which determines the proportion of maximal available demand ω_i. This demand function has an exponential form:

$$q(U) = 1 - \exp(-\lambda U) \tag{2.34}$$

with λ indicating the elasticity of demand.

The utility of customer at i from facility at j is measured by

$$u_{ij} = \sum_{r=1}^{R} k_{ijr} x_{jr} \tag{2.35}$$

Increases in the utility can be achieved by increasing a number of facility's attribute. To measure the utility of the customer at i from facility j an MSI-type model is used

$$u_{ij} = A_j(d_{ij} + 1)^{-\beta} \tag{2.36}$$

where A_j is the attraction felt by customers from facility j.

$$A_j = I\{j\}a_j \prod_{k=1}^{K} (1 + y_{jk})^{\theta_k} \tag{2.37}$$

K is the number of attractiveness attributes, a_j is the attractiveness level of the basic design, y_{jk} is the level of improvement over the basic level with respect to attribute k, and $I\{j\}$ an indicator function which is equal to 1 if facility opens at j and otherwise.

In the discrete case of the model only a finite number R of possible facility design is available. For a particular scenario $r \in R$, a_{jr} represents the attractiveness of facility located at j with design r, $j \in S$ where S represents the set of locations controlled by the decision maker. If

$$x_{jr} \begin{cases} 1 & \text{if facility with design scenario } r \text{ is locates at } j \\ 0 & \text{otherwise} \end{cases}$$

then the attractiveness A_j can be re-written as

$$A_j = \sum_{r=1}^{R} a_{jr} x_{jr}, j \in S \tag{2.38}$$

and the utility u_{ij} as

$$u_{ij} = \sum_{r=1}^{R} k_{ijr} x_{jr}, i \in N, j \in S \tag{2.39}$$

where $kijr = a_{jr}(d_{ij} + 1)^{-\beta}$ is a constant.

Let $U_i = \sum_{j \in S \cup C} u_{ij}$ denote the total utility of customers at i and $U_i(C) = \sum_{j \in C} u_{ij}$ the total utility of customers at i from the facilities controlled by the competitor (C is the set of locations controlled by the competitor) then the demand function at i will be

$$q(U_i) = 1 - \exp\left(-\lambda_i \left(\sum_{j \in S} \sum_{r=1}^{R} k_{ijk} x_{jr} + U_i(C)\right)\right) \tag{2.40}$$

and the total market share of the decision maker at i is given by

$$MS_i = \frac{\sum_{j\in S}\sum_{r=1}^{R} k_{ijr}x_{jr}}{\sum_{j\in S}\sum_{r=1}^{R} x_{ijr}x_{jr}} \tag{2.41}$$

Consequently the problem of the decision maker can be formulated as

$$\max \sum_{i\in N} \omega_i q(U_i))MS_i \tag{2.42}$$

$$st \quad \sum_{j\in S}\sum_{r=1}^{R} c_{jr}x_{jr} \leq B \tag{2.43}$$

$$\sum_{r=1}^{R} x_{jr} \leq 1, j \in S \tag{2.44}$$

$$x_{jr} \in \{0, 1\}, j \in S, r = 1, \ldots, R \tag{2.45}$$

For the solution of the discrete competitive facility and design problem an adapted weighted greedy heuristic algorithm is proposed.

Chapter 3
Sequential Competitive Facility Location Problems

3.1 Deterministic Single Level Sequential Problems

The formalization of this class of problem and the fundamental complexity results were established by Hakimi [28]. Following the game introduced by von Stackelberg [76], Hakimi [28] presented the two basic problems in sequential location analysis, the *centroid* and *medianoid* problems. These two problems are faced by the leader and the follower, respectively. The leader attempts to locate $p(\geq 1)$ facilities knowing that a follower will in turn locate his $r(\geq 1)$ facilities based on the leader's chosen locations; this is the $(r|p)$-centroid problem. The follower knows the set X_p that indicates where the leader's facilities are located, and solves an $(r|X_p)$-medianoid problem. Customers choose among the facilities according to a function of the distance between themselves and the facilities, preferring always the closest. This is the so-called binary customer choice. The formulation of the problems is based on the assumption that co-location is not allowed and if, by any chance the distance from a customer to the closest facility of the two competitors is the same, the customer always prefers the leader's facility. The demand of the customer is also considered to be inelastic with respect to the distance traveled.

If X and Y denote the location occupied by the leader and the follower, respectively, $d(j, X)$ and $d(j, Y)$ the distance between customer j and his nearest facility from X and Y respectively, then customer j prefers Y over X if $d(j, Y) < d(j, X)$ and he prefers X over Y otherwise. If $J(Y \prec X)$ is the set of customers who prefer Y over X, then $W(Y \prec X) = \sum_{j \in J(Y \prec X)} w_j$ denotes the total weight of the customers who prefer Y over X.

For each X the follower's strategy is the set of other location Y that provides the maximal market share, $W^*(X)$ to him. This maximal market share is obtained by solving the following problem:

$$\max_{Y, |Y|=r} W(Y \prec X). \tag{3.1}$$

A. Karakitsiou, *Modeling Discrete Competitive Facility Location*,
SpringerBriefs in Optimization, DOI 10.1007/978-3-319-21341-5_3

The leader on the other hand is interested in maximizing his own market share. Thus, his optimal location strategy X^* is the one that minimizes the follower's market share. Therefore, the leader's maximal market share is obtained by solving the following problem:

$$\min_{X,|X|=p} \max_{Y,|Y|=r} W(Y \prec X). \tag{3.2}$$

Hakimi in [29] extended the initial formulation of the problem by considering elastic demand and different customer choice rules, apart from the binary choice rule, such as partially binary choice and the proportional preference choice of the customers. Under the partially binary choice the customer uses the closest facility from each firm. Under the proportional choice the customer proportionally distributes his demand among the operating facilities. He came up with six different scenarios and he stated several vertex optimality results. Particularly, he proved the existence of a nodal solution for the partially binary problem, under both inelastic and elastic demand. He proved that a nodal solution exists also in the proportional choice-elastic demand case only if the demand captured by the facilities is a linear function of the distance. Suárez-Vega et al. [73] extended these results to concave function of the demand capture. In [67] the authors used polynomial discretization to approach the (r, X_p)-medianoid problem, while in [72] the authors provided polynomial time algorithms for the discrete (r, p)-centroid problem on path and $(1, p)$-centroid on trees.

A single level formulation of the $(r|p)$ formulation of the problem has been proposed in [63]. If s_{ij} is a binary variable indicating whether the leader's facility nearest to the customer j is placed at location i, that is,

$$s_{ij} = \begin{cases} 1 \text{ if customer } j \text{ patrinizes the leader's facility } i, \\ 0 \text{ otherwise,} \end{cases}$$

W_f is the total follower's market share, F is the set of all possible follower's solution, $y_f \in F$ denotes a set of r facilities opened by the follower, the problem can be stated as

$$\min \quad W_f \tag{3.3}$$

s.t

$$\sum_{i \in I} x_i = p, \tag{3.4}$$

$$s_{ij} \leq x_i, i \in I, j \in J, \tag{3.5}$$

$$\sum_{i \in I} y_{ij} = 1, j \in J, \tag{3.6}$$

$$W_f \geq \sum_{j \in J} w_j - \sum_{j \in J} \sum_{i \in I_j(y_f)} w_j s_{ij}, y_f \in F \tag{3.7}$$

$$s_{ij}, x_i \in \{0, 1\}, i \in I, j \in J, \tag{3.8}$$

where $I_j(y_f) = \{i \in I | d_{ij} \leq \min d_{kj} | k \in y_f, j \in J\}$ is the set of location that allow the leader to maintain customer j if follower uses the strategy y_f.

Constraint (3.7) is composed of an exponential number of constraints since F is composed of $\binom{|I|}{r}$ elements.

The main weakness of this family of inequalities is the fact that some of them make the lower bound weaker if the leader uses the same strategy y_f that defines the inequality because of the no co-allocation criterion [63]. To overcome this obstacle a family of valid inequalities by lifting (3.7) is proposed in [63]. To formulate the new family of lifted inequalities a function $H : y_f \to I$ is defined. This function gives an alternative place for each facility opened by the follower if the original place has already been used by the leader. The new family of lifted inequalities is as follows:

$$W_f \leq \sum_{j \in J} \sum_{i \in I_j(y_f) \cup \tilde{I}_j(y_f)} w_j s_{ij}, \quad y_f \in F, \tag{3.9}$$

where

$$I_j(y_f) = \left\{ i \in I | y_{f_i} = 0 \text{ and } d_{ij} \leq \min_{k \in I}(d_{kj} | y_{f_k} = 1) \right\},$$

$$\tilde{I}(y_f) = \left\{ i \in I | y_{f_i} = 1 \text{ and } d_{ij} \leq \min \left\{ \min_{k \in I}(d_{kj} | y_{f_k} = 1), d_{H(i)j} \right\} \right\}, \quad j \in J.$$

For finding some violated cuts a separation problem is defined. Given a fractional solution $(\bar{W}_f, \bar{x}, \bar{s})$ that satisfies (3.4)–(3.6) and some of the constraints (3.9), the separation problem is to find the follower's strategy $y_f \in F$ and a corresponding function $H : y_f \to I$ that minimizes the market share of the leader. In order to obtain the lower bound of W_f an exact branch and cut algorithm is applied to the separation problem.

A particle swarm optimization (PSO) procedure for solving problem (3.2) is proposed by Rodríguez et al. [64]. In order to adapt the PSO strategy to the leader–follower problem two swarms, X and Y for the leader and the follower respectively, are considered. The position of particle in X is a set of p candidate locations for the leader. Similarly, the position of each particle in Y is a set of r candidate location for the follower. For each particle in X, its maximum market share is estimated taking into account the position of the particles in Y. On the other hand, given the position of particles in X, particles in Y evolve to improve the market share captured by the follower. Both swarms evolve according to a Jumping Particle Optimization procedure [12].

Another type of sequential competitive facility location problem is the pre-emptive capture problem (PRECAP) which was studied by Serra and ReVelle in [69]. The PRECAP model analyzes the situation of a spatial market where there are as yet no competitors. A firm L (the leader) wants to locate p servers but knows that after locating its facilities one or several competitors will enter the market with several facilities. The only information that firm L has is that there will be q competitor's facilities located in the future (firm F, the follower). Thus, firm L wishes to pre-empt firm F in its bit to capture market share to the maximum extent possible. The mathematical formulation of the PRECAP problem is similar to the MAXCAP problem except that (2.2) and (2.3) cannot be written in extensive form. In [70] the assumption of the known number of the facilities being located by firm L is relaxed, that is, the market capture of firm L depends not only on the location of firm F, but also on the number of facilities located by it. In this case the problem can be formulate as

$$\max Z^L = \min \sum_{k=1}^{q} \frac{1}{k} \left(\sum_{j \in J} w_j y_{jk}^F + \sum_{j \in J} \frac{w_j}{2} z_{jk} \right) \tag{3.10}$$

$$st\ y_{jk}^F \leq \sum_{i \in P_j} s_{ik}^F, \forall j \in J, k = 1, \ldots, q \tag{3.11}$$

$$z_{jk} \leq \sum_{i \in T_j} s_{ik}^F, \forall j \in J, k = 1, \ldots, q \tag{3.12}$$

$$y_{jk}^F + z_{jk} \leq 1, \forall j \in J, k = 1, \ldots, q \tag{3.13}$$

$$\sum_{i \in I} s_{ik}^F = k, k = 1, \ldots, q \tag{3.14}$$

$$y_{jk}^F, z_{jk}, s_{ik} \in \{0, 1\}, \forall j \in I, \forall a \in Jk = 1, \ldots, q, \tag{3.15}$$

where variables are defined as follows:

$$y_{jk}^F = \begin{cases} 1, \text{ if firm } F \text{ captures } j \text{ when locating } k \text{ facilities} \\ 0, \text{ otherwise} \end{cases}$$

$$z_{jk} = \begin{cases} 1, \text{ if node } j\text{'s capture is divided between } L \text{ and } F \text{ when } L \text{ locates } k \text{ facilities} \\ 0, \text{ otherwise} \end{cases}$$

$$s_{ik}^F = \begin{cases} 1, \text{ if firm } F \text{ locates a facility at } i \text{ when locating } k \text{ facilities} \\ 0, \text{ otherwise} \end{cases}$$

and the parameters are:

- J the set of the demand areas, $j \in J$
- I the set of potential location sites, $i \in I$

- w_j the demand at node j
- P_i denotes the set of sites that firm F would patronize if she opens a new facility there.
- T_i denotes the set of sites of firm F that are located closed with the currently patronized competitor's facility
- k number of facilities that firm F can locate
- q maximum number of competitors facilities.

A two-phase one-opt heuristic is proposed in [70] to solve problem (3.10)–(3.15). The first phase is dealing with the computation of an initial solution which is found by locating the p facilities of the leader using any method and by locating 1 to q firm's F facilities using the MAXCAP problem. In the second phase the leader tries to improve the solution by relocating its servers using a one-opt procedure.

The problem addressed in [58] originates from the maximal covering location problem introduced by Church and ReVelle [11]. The authors revisited the problem and formulated a number of MIP models MaxMin, MinRegret, and Stackelberg. The Stackelberg model is solved with respect to leaders problem who anticipates the follower's strategic response to her location decision aiming at maximizing her market share remaining after follower's entrance. Facilities may be opened at a finite number of potential sites $s \in S$. Each leader's facility requires a fixed investment cost f_s, and the leader is restricted by her total investment budget B. On the other hand, the follower is assumed to open a single facility. T denotes the set of the available site for the location of the follower's single facility. The demand $d_i > 0$ is known and located at several location $i \in I$. The demand of a customer may be served by at most one of the firms through an open facility that covers (i.e., is close enough to serve) that customer. Each customer makes his decision solely based on distance; that is, he prefers the closest open facility. The Stackelberg model presented by Plastria Vanhaverbeke in [58] is the following:

$$\max \sum_{i \in I} d_i x_i^L \tag{3.16}$$

$$st \quad \sum_{s \in S} f_s Ls \leq B \tag{3.17}$$

$$\sum_{t \in T} F_t = 1 \tag{3.18}$$

$$x_i^L + x_i^F \leq 1, i \in I \tag{3.19}$$

$$x_i^L \leq \sum_{s \in S_i} L_s, i \in I \tag{3.20}$$

$$x_i^F \leq \sum_{t \in T_i} F_t, i \in I \tag{3.21}$$

$$1 - x_i^F \leq 1 - F_t + \sum_{s \in S_{it}} L_s, i \in I, t \in T_i \tag{3.22}$$

$$1 - x_i^L \leq \left(1 - \sum_{t \in T_i \backsim s \in S_{it}} F_t - \sum_{t \notin t_i} F_t\right) + (1 - Ls)\begin{pmatrix} i \in I, s \in S_i \\ \exists t \in T_i \backsim s \in S_{it} \end{pmatrix} \tag{3.23}$$

$$1 - y_{it} \leq \sum_{s \in S_{it}} L_s, i \in I, t \in T_i \backsim S_{it} \notin \emptyset \tag{3.24}$$

$$L_s \leq 1 - y_{it}, i \in I, t \in T_i, s \in s_{it} \tag{3.25}$$

$$\sum_{i \backsim t \in T_i, S_{it} \notin \emptyset} d_i y_{it} + \sum_{i \backsim t \in T_i, S_{it} \notin \emptyset} d_i \leq \sum_{i \in I} d_i x_i^F, t \in T \tag{3.26}$$

$$L_s, F_t, x_i^L, x_i^F \in \{0, 1\}, i \in I, s \in S, t \in T \tag{3.27}$$

$$y_{it} \in \{0, 1\} i \in I, t \in T \backsim S_{it} \neq \emptyset \tag{3.28}$$

where

- L_s 1 if the leader opens a facility at s, 0 otherwise
- x_i^L 1 if customer at i is captured by the leader, 0 otherwise
- F_t 1 if the follower opens a facility at t, 0 otherwise
- x_i^F 1 if customer at i is captured by the followers, 0 otherwise.
- y_{it} 1 if customers at i is captured by the followers when he locates at t
- S_i the capturing set of the leader, $S_i = \{s \in S; \delta_i^s \leq \delta_i$, and T_i the capturing set of the follower, $T_i = t \in T; \delta_i^t \leq \delta_i$ where δ_i is the distance from the customer i to the current nearest competitor already present on the market and δ_i^s, (δ_i^t) the distance between costumer at i and site s, (t).
- S_{it} the set of sites of S that i prefers to t, $S_{it} = \delta_i^s < \delta_i^t$

3.2 Probabilistic Single Level Sequential Problems

In the work of Suárez-Vega et al. [74] is presented a generalization of the $(r|X_p)$-Medianoid problem where the choices of customers are based not only on the distance but on certain facilities, characteristics. The follower aims at finding not only the optimal place for his facilities but also their attractiveness. The attractiveness is incorporated into the model by using a Huff type attraction function [34]. Given $Y_r = (y_1, y_2, \ldots, y_r)$ and $X_p = (x_1, x_2, \ldots, x_p)$, the locations of the facilities belonging to the follower firm, F, and the leader, L, respectively, with attractiveness levels $A_r = (a_1, a_2, \ldots, a_r)$ and $A_p = (b_1, b_2, \ldots, b_p)$, the profit function for follower is formulated as

$$W(Y_r, A_r|X_p, A_p) = \sum_{v \in V} w_Y(v) - \sum_{j=1}^{r} F(a_j) \tag{3.29}$$

with $w_Y(v)$ the demand at node v which is captured by F.

The solution of the problem

$$W(Y_r^*, A_r^*|X_p, A_p) = \max_{Y_r \in N_r, A_r \in [I,S]^r} W(Y_r, A_r|X_p, A_p) \quad (3.30)$$

constitutes a $(r|X_p, A_p)$-Medianoid.

This problem has been analyzed under different customer choice rules, binary, partially binary, and proportional preferences. The authors proved that for binary preferences, a discretization of the location problem can be applied if the attractiveness levels are given. To solve the discrete problem three combinatorial heuristics combined with a global search procedure based on branch and bound techniques were used. The optimum attractiveness levels can be found evaluating, at most, $(|V| + 1)^r$ candidates.

3.3 Deterministic Multilevel Sequential Problems

A bi-level formulation of the game presented by Hakimi [28] is proposed by Allekseeva et al. [3]. For a solution x used by the leader, Alekseeva et al. [3] define the set of facilities which allows the follower to capture customer j as:

$$J_j(x) = \{i \in I | d_{ij} < \min_{l \in I | x_l = 1} d_{lj}\}, j \in J$$

By introducing variables

$$x_i = \begin{cases} 1 & \text{if facility } i \text{ is opened by leader,} \\ 0 & \text{otherwise,} \end{cases} \quad (3.31)$$

$$y_i = \begin{cases} 1 & \text{if facility } i \text{ is opened by follower,} \\ 0 & \text{otherwise,} \end{cases} \quad (3.32)$$

$$z_j = \begin{cases} 1 & \text{if customer } j \text{ is served by leader,} \\ 0 & \text{otherwise,} \end{cases} \quad (3.33)$$

a linear IP bi-level programming problem of the game can be stated as [3]:

$$\max_x \textstyle\sum_{i \in J} w_j z_j^*(x) \quad (3.34)$$

$$st. \quad \sum_{i \in I} x_i = p, \quad (3.35)$$

$$x_i \in \{0, 1\}, \forall i \in I \quad (3.36)$$

where $z_j^*(x)$ solves

$$\max_{y,z} \sum_{i \in J} w_j(1 - z_j) \tag{3.37}$$

$$st. \qquad \sum_{i \in I} y_i = r \tag{3.38}$$

$$1 - z_j \leq \sum_{i \in I_j(x)} y_i, j \in J, \tag{3.39}$$

$$x_i + y_i \leq 1, i \in I \tag{3.40}$$

$$y_i, z_i \in \{0, 1\}, i \in I, j \in J. \tag{3.41}$$

The objective function (3.34) defines the market share of the leader. Constraint (3.35) ensures that the leader opens exactly p facilities. The objective function (3.37) defines the market share of the follower. Constraint (3.38) ensures that the leader opens exactly r facilities. Constraints (3.39) state that if the follower has no facility in the set $I_j(x)$, then customer j is served by the leader. Finally, constraints (3.40) guarantee that each facility can be opened by at most one decision maker.

In order to solve the problem Alekseeva et al. [3] have proposed a hybrid memetic algorithm, where a Tabu search approach is used to improve the elements of population.

An upper bound is obtained by reformulating the bi-level problem as a single level mixed integer problem with an exponential number of constraints and variables. If $\mathscr{F}$ is a family of follower solutions and $I_j(y) = \{i \in I | d_ij \leq \min_{l \in I}(d_{lj} | y_l = 1)\}, y \in \mathscr{F}, j \in J$ is the set of facilities which allow the leader to keep the customer j if the follower uses solution y, and if $\mathscr{F}$ contains all possible solutions of the follower then the problem (3.34)–(3.41) is equivalent to the following $0-1$ program:

$$\max \quad W \tag{3.42}$$

$$st \quad \sum_{j \in J} w_j x_{iy} \geq W, y \in \mathscr{F}, \tag{3.43}$$

$$z_{iy} \leq \sum_{i \in J_j(y)} x_i, j \in J y \in \mathscr{F}, \tag{3.44}$$

$$\sum_{i \in I} x_i = p, \tag{3.45}$$

$$x_i, z_{iy} \in \{0, 1\}, i \in I, j \in J, y \in \mathscr{F} \tag{3.46}$$

where $W \geq 0$ is the total market share of the leader and z_{iy} is binary variable indicating whether customer j is serviced by the leader when the follower uses a solution y. In order to get the appropriate family the Probabilistic Tabu Search algorithm is used. This single level model is also used to find the global optimum. An iterative exact algorithm is developed for this purpose.

The authors in [13], based on the bi-level representation (3.34)–(3.41), approached the leader's–follower's problem using two metaheuristic methods: local search with variable neighborhoods and stochastic Tabu search. In order to apply the variable neighborhood method, they introduced three neighborhoods to solve x: (a) Neighborhood $Fswap(x)$ consists of all leader's solutions that result from x by closing one facility and opening one facility from the solution $Y^*(x)$. (b) Neighborhood $Nswap(x)$ consists of all leader's solutions that result from x by closing one facility and opening another, located at no more than the lth facility nearest to it. (c) Neighborhood $Cswap(x)$ is the complement of these two neighborhoods, $Cswap(x) = Swap(x) \backslash Fswap(x) \cup Nswap(x)$. On each iteration, a probabilistic procedure $Shake(x, k)$ is applied to the leader's current solution x in order to replace k randomly chosen facilities of the leader by other randomly chosen facilities. Then, a local improvement procedure first with respect to neighborhoods $Fswap$ and $Nswap$, and, if necessary, with respect to the neighborhood $Cswap$ is applied to the resulting solution x'. The current local optimum x'' is compared to the solution x. If the new local optimum is better than the previous one, it is passed to the new local optimum. In order to apply the local tabu search a part of the $Swap(x)$, $Swap_q(x)$ is chosen at random. Each element of the sets $Swap(x)$ is included in the set $Swap_q(x)$ independently of other elements with nonzero probability q.

A three-level programming formulation for both leader's and followers' problem and a heuristic solution procedure based on the elimination procedure in a candidate list are proposed in [10]. They formulated the problem as a three-stage optimization process which included the customer selection problem, the follower location set problem, and the leader location set problem. The corresponding $(r|p)$-centroid problem with inelastic demand is as follows:

$$\max \sum_{i=1}^{m} \left[\sum_{k=1}^{n} h_k z_{ki} \right] x_i \tag{3.47}$$

$$s.t \sum_{i=1}^{m} x_i = p \tag{3.48}$$

$$x_i \in \{0, 1\}, i \in [1, \dots, m], \tag{3.49}$$

where z solves

$$CUS(x, y) \min \sum_{k=1}^{n} \sum_{i=1}^{m} d_{ki} h_k z_{ki} \tag{3.50}$$

$$st \sum_{i=1}^{m} z_{ki} = 1, k \in [1, \dots, n] \tag{3.51}$$

$$z_{ki} \leq \bar{x}_i + \bar{y}_i, k \in [1, \ldots, n], i \in [1, \ldots, m] \tag{3.52}$$

$$z_{ki} \in \{0, 1\}, k \in [1, \ldots, n], i \in [1, \ldots, m], \tag{3.53}$$

where y solves

$$FLO_r \max \sum_{i=1}^{m} \sum_{k=1}^{n} h_k z_{ik} \tag{3.54}$$

$$st \sum_{i=1}^{m} y_i = r, \tag{3.55}$$

$$\sum_{i=1}^{m} z_{ki} \leq 1, k \in [1, \ldots, n] \tag{3.56}$$

$$z_{ki} - c_{ki} y_i \leq 0, k \in [1, \ldots, n], i \in [1, \ldots, m] \tag{3.57}$$

$$z_{ki}, y_i \in \{0, 1\}, k \in [1, \ldots, n], i \in [1, \ldots, m], \tag{3.58}$$

where m is the number of possible facility locations and n is the number of customer locations. $d_{ki} = d(c_k, f_i)$ is the distance between the kth customer location c_k and the ith facility point f_i and h_k is the total demand of the customers located at c_k. A set Z of location points is identified by a binary vector $z = (z_i : i \in [1, \ldots, m])$ where $z_i = 1$ if $f_i \in Z$ and $z_i = 0$ if $f_i \notin Z$. The decision variables in the leader and follower location problems are the m-vectors x and y of binary decision variables corresponding to sets X and Y the location sets for the leader and the follower, respectively. z_{ki} are the $0 - 1$ decision variables indicating whether the customers located at the k customer location c_k prefer the location f_i for the facility. While coefficient c_{ki} defines the set of customers captured by each single follower location, and $c_{ki} = 1$ if $d_{ki} < \min\{d_{kj}\bar{x}_j = 1; 0 \text{ otherwise}\}$.

The $CUS(x, y)$ is the customer selection problem. The objective function (3.50) of this problem represents the total distance traveled by the customers to arrive at the corresponding facility points. The constraints (3.51)–(3.52) state that each customer has to go to one location in the leader location set or in the follower location set. FLO_r corresponds to the follower's location set problem, respectively. Given the leader locations $\bar{x}$, this problem consists of selecting the r locations y that capture the largest amount of customer demand.

Based on the observation that the mathematical programming formulation of the minimax problem that corresponds the leader's problem (3.47)–(3.58) is

$$\max \ W \tag{3.59}$$

$$st \ |X| = p \tag{3.60}$$

$$W(Y \prec X) \leq W, \forall Y \in L^r, \tag{3.61}$$

Campos Rondríguez et al. [10] proposed a heuristic based on an elimination procedure in a candidate list in order to solve the leaders problem. In the procedure, a leader solution provides an upper bound for the leader–follower problem. A family F of good follower candidates is used to conclude that the upper bound provided by a leader solution cannot be improved, and therefore, this solution is an optimal solution.

3.4 Probabilistic Bi-level Sequential Problems

The aim of the model proposed in [44] is to determine the optimal location and the attractiveness of the new facilities to be opened by a firm in a market where there are already r existing facilities that belong to one or more competitors. The goal is the maximization of the firm's profit. The customers are aggregated at $N = 1, \ldots, n$ demand points and the number of candidate facility site is $M = 1, \ldots, m$. The parameters of the problem are

- a_j annual buying power at point j
- c_i unit attractiveness cost at site i
- f_i annualized fixed cost of opening and operating a facility at i
- d_{ij} Euclidean distance between site i and point j
- b_j total utility of existing facility depending on their attractiveness and distance from point j
- u_i maximum attractiveness level of facility to be opened at site i
- q_k attractiveness of existing facility j

and variables

- Q_i attractiveness of the facility opened at site i
- X_i binary variable that is equal to 1 if a facility is opened at site i and 0 otherwise.

By using Huff's model [34], the utility of the facility opened at site i with attractiveness Q_i is defined by Q_i/d_{ij}^2. By using the same rule the total utility felt by customers at j for the existing facilities is $b_j = \sum_{k=1}^{r} q_k/d_{kj}^2$, where d_{kj} is the distance between demand point j and existing facility k. Hence, the market share of the facility at i is expressed as

$$P_{ij} = \frac{Q_i/d_{ij}^2}{\sum_{i=1}^{m}(Q_i/d_{ij}^2) + \sum_{k=1}^{r} q_k/d_{kj}^2} \tag{3.62}$$

As a result the total revenue captured by the new facility is given by

$$\sum_{j=1}^{n} a_j \frac{\sum_{i=1}^{m}(Q_i/d_{ij}^2)}{\sum_{i=1}^{m}(Q_i/d_{ij}^2) + \sum_{k=1}^{r}(q_k/d_{kj}^2)} \tag{3.63}$$

Then the problem can be formulated as

$$\max_{\mathbf{Q},\mathbf{X}} z = \sum_{j=1}^{n} a_j \frac{\sum_{i=1}^{m}(Q_i/d_{ij}^2)}{\sum_{i=1}^{m}(Q_i/d_{ij}^2) + \sum_{k=1}^{r}(q_k/d_{kj}^2)}$$

$$- \sum_{i=1}^{m} f_i X_i - \sum_{i=1}^{m} c_i Q_1 \tag{3.64}$$

$$s.t\ Q_i \leq u_i X_i, i = 1, \ldots, m \tag{3.65}$$

$$X_i \in \{0, 1\}, i = 1, \ldots, m \tag{3.66}$$

$$Q_i \geq 0, i = 1, \ldots, m \tag{3.67}$$

To solve the problem three solution methods are presented. One is a heuristic based on the Lagrangian relaxation of the model, while the other two are exact procedures based on the branch and bound technique.

The model proposed in [45] allows the competitor to react in every location decision made by the firm by adjusting the attractiveness level of his own existing facilities with the objective to maximize his profit. The resulting formulation is a bi-level programming model where the entering firm is considered as the leader and the existing competitor as the follower. Notice that in this bi-level formulation, the attractiveness level at the competitor's facility q_k becomes the decision variable of the follower.

Thus, the leader solves problem (3.64)–(3.67), while the follower solves the problem

$$\max_{\mathbf{q}} \sum_{j=1}^{n} a_j \frac{\sum_{i=1}^{r}(q_k/d_{kj}^2)}{\sum_{i=1}^{m}(Q_i/d_{ij}^2) + \sum_{k=1}^{r}(q_k/d_{kj}^2)} -$$

$$-\sum_{k=1}^{r} \tilde{c}_k (q_k - \tilde{q}_k), \tag{3.68}$$

$$s.t.\ \ q_k \leq \bar{q}_k, k = 1, \ldots, r \tag{3.69}$$

$$q_k \leq 0, k = 1, \ldots, r, \tag{3.70}$$

where the first term of the objective function represents the follower's market share, and $\tilde{q}_k, \bar{q}_k, \tilde{c}_k$ are parameters representing the current attractiveness level, the maximum attractiveness level, and the unit attractiveness cost of the competitor's facility k, respectively. The authors prove the concavity of the follower' objective function with respect to attractiveness level q. Making use of this property the authors transform the bi-level model into an equivalent single level model mixed integer nonlinear program so that it can be solved by global optimization methods. The transformation is done by substituting the KKT first order conditions into the leader's problem.

The model was further developed in [46] so as to allow the follower to make decisions not only regarding the attractiveness level but also regarding location.

The bi-level models proposed by Beresnev [5] assumed that each client selects a serving facility on the basis of his own preferences, which can rank (order) all facilities available to be opened.

The author [5] considers two settings of the problem that differ in the objective functions of the follower firm: In the first, it is assumed that the goal of the leader firm as well as the follower firm is the maximization of the profit, while in the second, the objective of the follower is maximization of his income, it is also assumed that each facility opened by the follower firm cannot be loss-making. The author uses the following notation in order to build up the proposed models

- $I = \{1, \ldots, m\}$ is the set of possible sites for location;
- $J = \{1, \ldots, n\}$ is the set of clients;
- p_{ij} is the income realized by facility $i \in I$ open by leader when serving client $j \in J$;
- q_{ij} is the income realized by facility i opened by the follower firm when serving client $j \in J$;
- $\prec_j$ is a linear order on I determining the preferences of client $j \in J$, and $i \prec_j k$ means that of the two open facilities i and $k \in I$ client j selects facility i; the relation $i \preceq_j k$ means that either $i \prec_j k$ or $i = k$;
- f_i is the fixed cost of the leader firm for opening facility $i \in I$;
- g_i is the fixed cost of the follower firm for opening facility $i \in I$.
- x_i is the variable indicating if facility $i \in I$ is opened by the leader firm,
- x_{ij} is the variable indicating if facility $i \in I$ opened by the leader firm is selected by client $j \in J$;
- z_i is the variable indicating if the follower firm opens facility $i \in I$;
- z_{ij} is the variable indicating if client $j \in J$ selects facility $i \in I$ opened by the follower firm

When the goal of the follower firm is to maximize the profit, the model is written as follows:

$$\max_{(x_i),(x_{ij})} \left\{ -\sum\nolimits_{i \in I} f_i x_i + \sum_{j \in J} \left(\sum_{i \in I} p_{ij} x_{ij} \right) \left(1 - \sum_{i \in I} \tilde{z}_{ij} \right) \right\} \tag{3.71}$$

$$st \qquad \sum_{i \in I} x_{ij} = 1, j \in J \tag{3.72}$$

$$x_i \geq x_{lj}, i \in I, j \in J, \tag{3.73}$$

$$x_i + \sum_{i \prec_j l} x_{ij} \leq 1, i \in I, j \in J \tag{3.74}$$

$$x_i, x_{ij} \in \{0, 1\}, i \in I, j \in J \tag{3.75}$$

where $(\tilde{z}_i)$, $(\tilde{z}_{ij})$ is the optimal solution of the following problem:

$$\max_{(z_i),(z_{ij})} \left\{ -g_i z_i + \sum_{j \in J} \sum_{i \in I} q_{ij} z_{ij} \right\} \tag{3.76}$$

$$st \quad \sum_{i \in I} z_{ij} \leq 1, j \in J \tag{3.77}$$

$$z_i \geq z_{ij}, i \in I, j \in J \tag{3.78}$$

$$x_i + z_i + \sum_{i \prec_j l} z_{lj} \leq 1, i \in I, j \in J \tag{3.79}$$

$$z_i, z_{ij} \in \{0, 1\} i \in I, j \in J \tag{3.80}$$

Objective function (3.71) shows the value of profit received by the leader taking into account that a part of his consumers will be captured by the follower. Constraint (3.72) guarantees that each client can select one facility from the leader and inequalities (3.73) that only one open facility can be selected. Inequalities (3.74) implement the rule for choosing a facility opened by the Leader to service a consumer. The same inequalities guarantee that to service each consumer one can choose only one facility opened by the Leader. Objective function (3.76) of problem shows the value of the profit received by the follower. Inequalities (3.79) implement conditions for the follower capturing consumers for given facilities opened by the Leader.

The computational complexity of problem (3.71)–(3.80) is discussed in [49], where it is proved that the problem is Σ_2^p-hard when the cost of opening facilities is considered as null.

The method proposed in [5] for computing the upper bound for the objective function of the problems (3.71)–(3.80) includes the two phases: In the first phase, the estimation pseudo-Boolean function is constructed and, in the second phase, its minimum value is computed.

A number of solution methods of the problem have been proposed in a series of publication [6–8, 50]. Their main characteristic is that they are based on the maximization of a pseudo-Boolean function of the form

$$\max_x \quad f(x) \tag{3.81}$$

$$st \ x \in B^m. \tag{3.82}$$

3.5 Bi-level Models with Pricing Decisions

Models for duopolistic competitive location planning with sequential acting and variable delivered price are introduced in [22]. To formulate the problems Fischer [22] considered the following two competitors situation: Both competing firms sell a homogeneous product to their customers aiming at maximizing their profit.

The participating firms are supposed to be rational and to have perfect foresight as well. The resulting two-stage game can be described as: The first decision maker (leader) chooses her strategy anticipating the optimal decision of the competitor. The second decision maker (the follower) is able to react in an optimum way to every decision of the leader, i.e., to determine a reaction function,

The number of locations the competitors want to establish is fixed p for the leader and r for the follower. The choice of one or more identical locations by both competitors is supposed to be possible. However it is shown that identical location and sharing of markets are never an optimal choice under the considered circumstances.

The demand of the customers occurs at a set J of discrete points called markets, and is assumed to be elastic and depends on price $p_j, j \in J$, so the demand at each market is given by the demand function $D_j(p_j)$. As a results firms' decisions include not only locations from a set of potential locations I but also delivered price p_{ij} settings at the demand markets. The pricing policy utilized relies on the "discriminatory pricing policy." In addition, it is assumed that the demand in each market has to be fulfilled completely.

The transportation cost per unit of product and unit of distance is denoted by t and the distance between location i and market j is known to be d_{ij}. For simplicity it is assumed that production cost is zero.

Due to homogeneity assumption customers always buy from the competitor who charges the lowest price, and do not have any other preferences concerning their suppliers. However there is a limit in price $p_{j\max}$ above which demand decreases to zero $(D_j(p_{j\max}) = 0)$. On the other hand, the maximum demand $D_{j,\max}$ occurs if nothing is charged for the product, i.e. $D_j(0) = D_{j,\max}$.

If both players choose the location and price "once and for all," the optimal location and price decision can be found by solving the following problem:

$$\max_{z,v,p_{ij}^L} \sum_{i \in I} \sum_{j \in J} D_j(p_{ij}^L) z_{ij} \left(p(L)_{ij} - td_{ij}\right) \tag{3.83}$$

$$st \quad v_i \geq z_{ij}, \forall i \in I, \forall j \in J \tag{3.84}$$

$$\sum_{i \in I} v_i = p \tag{3.85}$$

$$\sum_{i \in I} z_{ij} \leq 1, \forall j \in J, \tag{3.86}$$

where $x, v, P(F)_{ij}$, being the solution of

$$\max_{x,y,p_{ij}^F} \sum_{i \in I} \sum_{j \in J} D_j(p_{ij}^F) x_{ij} \left(p_{ij}^F - td_{ij}\right) \tag{3.87}$$

$$st \quad y_i \leq x_{ij}, \forall i \in I, \forall j \in J \tag{3.88}$$

$$\sum_{i \in I} y_i = r, \tag{3.89}$$

$$\sum_{i \in I} z_{ij} + x_{ij} = 1, \forall j \in J \tag{3.90}$$

$$\left(P_{ij}^F - p_{ik}^L\right) - \left(z_{ik} - x_{ij}\right) \leq 0, \forall k, i \in I, \forall j \in J \tag{3.91}$$

$$x_{ij}, z_{ij} \in \{0, 1\}, \forall i \in I, \forall j \in J \tag{3.92}$$

$$v_i, y_i \in \{0, 1\}, \forall i \in I, \tag{3.93}$$

$$p_{ij}^L \geq 0, p_{ij}^F \geq 0, \forall i \in I, \forall j \in J, \tag{3.94}$$

where

- z_{ij} binary variable indicating whether leader serves market j from location i.
- x_{ij} binary variable indicating whether follower serves market j from location i.
- v_i binary variable, takes value 1 if leader opens a facility at location i.
- y_i binary variable, takes value 1 if follower opens a facility at location i.
- p_{ij}^L continuous variable, gives the price set by the leader at market j if he serves it from location i.
- p_{ij}^F continuous variable, gives the price set by the follower at market j if he serves it from location i.

Objective function (3.83) and (3.87) maximize the profit of the players. Conditions (3.84) and (3.88) make sure that demands can only be served by a player from a certain location i if a facility has been opened there. By conditions (3.85) and (3.85) the number of each competitors is fixed to the planned number. Constraints (3.86) and (3.90) make sure that each market is served exactly one. Constraints (3.91) make sure that market j is served from the cheapest location.

However when price adjustment is possible, another stage has to be added to the model. The aim of this pricing stage is to search for a Nash equilibrium in prices. For each combination of a leader's and a follower's location the optimal prices are calculated as follows: Let i^* be the location of the leader which is nearest to market j and let k^* be the analogous location of the follower. Then follower can capture the demand of market j if

$$p_{k^*j,\min}^F = td_{k^*j} \leq p_{i^*j,\min}^L = td_{i^*j}$$

and the optimal price for this location combination can be found solving the problem:

$$\max D_j(p_{i^*jk^*}^F)(p_{i^*jk^*}^F - td_{ij}) \tag{3.95}$$

$$st \quad p_{i^*j,k^*}^F \in \left[\min\left\{p_{k^*j,\min}^F; p_{i,\max}\right\}, \min\left\{p_{i^*j,\min}^L; p_{i,\max}\right\}\right] \tag{3.96}$$

where $\left[\min\left\{p^F_{k^*j,\min}; p_{i,\max}\right\}, \min\left\{p^L_{i^*j,\min}; p_{i,\max}\right\}\right]$ is the interval of rational pricing strategies for follower. The same procedure can be applied for the leader since he can adjust his price after the location of both players become common knowledge. This third Nash game stage is incorporated into the objective function coefficients using auxiliary variables

- x_{ijk} indicates whether market j is served by follower from location k while i is the location of the leader which is nearest to $j (i \neq k)$,
- z_{ijk} indicates whether market j is served by the leader from location i while k is location of the follower nearest to $j (i \neq k)$.

The resulting model is

$$\max_v \sum_{i \in I} \sum_{j \in J} \sum_{k \in I} D_j(p^L ijk) - (p^L_{ijk} - td_{ij}) z_{ijk} \tag{3.97}$$

$$st \quad (3.85), \tag{3.98}$$

where x, y, z being the solution of

$$\max_y \sum_{i \in I} \sum_{j \in J} \sum_{k \in I} D_j(p^F_{ijk})(p^F_{ijk} - td_{ik}) x_{ijk} \tag{3.99}$$

$$st \quad v_i \leq \sum_{k \in I} x_{ijk}, v_i \leq \sum_{k \in I}, \forall i \in I, \forall j \in J \tag{3.100}$$

$$y_k \leq \sum_{i \in I} x_{ijk}, y_k \leq \sum_{i \in I} z_{ijk}, \forall j \in J, k \in I, \tag{3.101}$$

$$(3.89), \tag{3.102}$$

$$\sum_{i \in I} \sum_{k \in I} (z_{ijk} + x_{ijk} = 1, \forall j \in J \tag{3.103}$$

$$p^F_{ijk} - p^L_{hjo}(z_{hjo} - x_{hjo} \leq 0, \forall j \in Jh, i, k, o \in I \tag{3.104}$$

$$x_{ijk}, z_{ijk} \in \{0, 1\}, \forall j \in J, i, k \in I \tag{3.105}$$

$$v_i, y_i \in \{0, 1\}, \forall i \in I. \tag{3.106}$$

The heuristic procedure from problem (3.99)–(3.106) presented by Fischer [22] strongly relies on the determination of the follower's reaction function which is given by the binary model which is equivalent to the maximum capture problem [60]:

$$\max \sum_{k \in I} \sum_{j \in J} P_{kj} x_{kj} \tag{3.107}$$

$$st \quad (3.88), (3.89)$$

$$\sum_{i \in I} x_{ij} = 1, \forall j \in J \tag{3.108}$$

$$x_{ij}, y_i \in \{0, 1\}, \forall i \in I, \forall j \in J, \tag{3.109}$$

where P_{kj} is the follower's profit resulting from serving market j from location k, when the leader location is known, that is, the location in which the inequality $p^F_{ijk} < p^L_{ijk}$ holds.

Thereafter, the procedure is analogous to the PRIMAL algorithm presented in [71]. The only difference is that in this case the objective function of both players has to be calculated individually, and in addition at least one of the objective function coefficients must be recalculated at each iteration.

Chapter 4
Simultaneous Competitive Facility Location Models

The research effort of this body of literature aims at developing insights concerning the equilibrium pattern of location decisions and their stability. Another subject that is addressed by several authors is the existence or (not) of a set of locations and pricing or production quantities that will ensure a Nash equilibrium, that is, a position where neither firms have incentives to move.

4.1 Stability in Location Decisions

In 1987, Dobson and Karmarkar [14] stated the notion of stability in the facility location decision. A set of facilities is considered stable if each facility is economically viable and no competitor can successfully open any facilities. The problem was formulated as a three-stage optimization process which included the choice of the existing firm, the competitor, and the customers. The purpose of the analysis was to identify the maximum-profit set of locations that is stable under competition. If y_j is a binary variable indicating whether the existing firm has open location j, z_j is the respective re-binary variables for the competitor and x_{ij} is also binary variable indicating whether the customers at location i travel to location j, then for a given set of open locations (y, z), the customers' problem is

$$\min_{x_{ij}} \sum_{i,j} t_{ij} d_i x_{ij} \tag{4.1}$$

$$st \sum_{j} x_{ij} = 1, \forall i \tag{4.2}$$

$$x_{ij} \leq \bar{y}_j + \bar{z}_j, \forall i, j \tag{4.3}$$

$$x_{ij} \in \{0, 1\} \forall i, j, \tag{4.4}$$

A. Karakitsiou, *Modeling Discrete Competitive Facility Location*, SpringerBriefs in Optimization, DOI 10.1007/978-3-319-21341-5_4

where t_{ij} is the cost of travel between i and j by the cheapest route and d_i is the demand from customers at location i. If further p denotes the price at all locations, c_j is the variable cost of satisfying demand from a facility at location j and K_j is the fixed cost of opening a facility at location j then the competitors problem, given a set of open location owned by the existing firm (y), is

$$\max_{z_j} \sum_j \left[(p - c_j) \left(\sum_i d_i x_{ij} \right) - K_j \right] z_j \tag{4.5}$$

$$st \qquad \bar{y}_j + z_j \leq 1, \forall j \tag{4.6}$$

$$z_j \in \{0, 1\}, \forall j \tag{4.7}$$

where x solves (4.1)–(4.4), and the existing firm's problem is

$$\max_{y_j} \sum_j \left[(p - c_j) \left(\sum_i d_i x_{ij} \right) - K_j \right] y_j \tag{4.8}$$

$$z_j \in \{0, 1\}, \forall j, \tag{4.9}$$

where x solves (4.1)–(4.4) and z solves (4.5)–(4.7). The models were solved using enumeration algorithms.

4.2 Location-Pricing Equilibria

The location problems addressed by this type of location equilibrium analysis assume that the competing firms are dealing with a Bertrand game type in their effort to maximize profits, that is, apart from location they decide also on the price to be set in each market. In location literature, firms typically use a mill pricing policy or a delivered pricing policy. The seller sets the same price to all customers in the market and the buyer pays the transportation cost, in the first, while the seller charges a specific price in each market and takes care of the transportation cost in the second.

The main characteristic of location-price models is the lack of equilibrium with simultaneous decisions on location and price [25]. This obstacle is usually overcome by employing a two-stage procedure in order to investigate whether or not such equilibria exist. This division is motivated by the fact that in real-life situations a firm first decides its location and then its pricing policy. More precisely, firms are supposed to choose location and price one at a time in a two-stage game, with the aim of maximizing their own profits. In the first stage firms simultaneously make their location decision. Given the outcome of the first stage, firms then simultaneously determine their price or production level in the second stage.

The solution concept used is so-called sub-game perfect Nash equilibrium, which captures the idea that when firms select their location, they all anticipate the consequences of their choice on price competition.

However, when a firm sets a mill pricing, a price equilibrium at the second stage of the game rarely exists. As a result the majority of the research work devoted to discrete or network location problems have been using parametric prices [25, 71].

On the other hand, the existence of a price equilibrium is possible when a delivered pricing mechanism is used. Lederer and Thisse [48] consider a duopoly with inelastic demand and constant marginal cost. Customers are located at $z_h, h = 1, \ldots, H$ distinct markets on a network $N = (V, E)$. At each market z_h customers have fixed demand of a total amount D_h. The unit transportation cost between firm's $i, i = (1, 2)$ location x and market z_h is denoted by $t_i = t_i[d(x_i, z_h)]$ and is assumed to be positive, concave and increasing with respect to distance. Firm i produces the product by using J inputs $j = 1, \ldots, J$. There are H_j possible sources of input j in N, denoted by $y_{jh}, h = 1, \ldots, H_j$. The price of input j, at y_{jh}, is given and denoted by p_{jh}. The cost of transporting one unit of input j from y_{jh} to x_i is $t_j(d(y_{jh}, x_i))$, which is assumed to be increasing and concave with distance $d(y_{jh}, x_i)$. If q_{ji} is the amount of input j used by firm i to produce one unit of the product then the marginal production cost at x_i is given by

$$m_i(s_i) = \sum_{j=1}^{J} \left[\min_{h=1,\ldots,H_j} \{p_{jh} + t_j[d(y_{jh}, x_i)]\} \right] q_{ji}$$

The firms compete through their decisions concerning location, production technology, and delivered prices. The problem is formulated as a two-stage game. In the first stage each firm i chooses a location x_i and a production technology τ_i. For a given location-technology choice s_i the most efficient mix of inputs to produce output $q_i > 0$ is the one that minimizes the firm's i total cost which is given by

$$TC_i(q_i, s_i) = F_i(s_i) + C_i(q_i, s_i), \tag{4.10}$$

where F_i and $C_i = m_i(s_i)q_i$ denote the fixed and variable production cost, respectively. In the second stage, the firms simultaneously set delivery price schedules. For the solution of the problem, Lederer and Thisse [48], first, they proved that for any location-technology choices (s_1, s_2) there exists a unique pair of delivered prices (p_1^*, p_2^*) satisfying,

$$p_1^*(z_h) = p_2^*(z_h) = \max\{m_1(s_1) + t_1[d(x_1, z_h)], m_2(s_2) + t_2[d(x_2, z_h)]\}$$

for all H markets, which constitutes a unique Nash equilibrium in price (stage two). Then, this equilibrium is used to search for the equilibrium in location and technology at the first stage of the game. The study of the equilibrium of the first stage is facilitated by the use of the so-called social cost function.

$$SK(s_1, s_2) = \sum_{h=1}^{H} \min\{m_i(s_i) + t_i[d(x_i, z_h)]\}D_h + F_1(s_1) + F_2(s_2) \tag{4.11}$$

The social cost is defined as the total delivered cost if each customer were served with the lowest marginal delivered cost. Lederer and Thisse [48] show that a location equilibrium exists and is the global optimizer of the social cost. As in Hakimi [28], they also proved that if the firm transport costs are strictly concave, then the set of locational choices of the firm is reduced to the vertices of the network.

Dorta-González et al. [15] obtained the same result in the case of oligopoly, i.e., when $1 \leq i \leq r$. They proved the existence of a subgame perfect Nash equilibrium at the vertices of the network when transportation costs are concave with respect to distance.

In Pelegrín-Pelegrín et al. [56] the case of multifacility duopoly is considered and the following integer linear programming formulation aiming at minimizing the social cost is proposed to find a location equilibrium in discrete location space.

$$\min \sum_{h=1}^{H} D_h \sum_{\ell \in L_1} p_{1h}^{\ell} y_{1h}^{\ell} + \sum_{h=1}^{H} D_h \sum_{\ell \in L_2} p_{2h}^{\ell} y_{2h}^{\ell} \tag{4.12}$$

$$s.t. \sum_{\ell \in L_1} y_{1h}^{\ell} + \sum_{\ell \in L_2} y_{2h}^{\ell} = 1, h \in H \tag{4.13}$$

$$y_{1h}^{\ell} \leq x_1^{\ell}, h \in H, \ell \in L_1 \tag{4.14}$$

$$y_{2h}^{\ell} \leq x_2^{\ell}, h \in H, \ell \in L_2 \tag{4.15}$$

$$\sum_{\ell \in L_1} x_1^{\ell} = r \tag{4.16}$$

$$\sum_{\ell \in L_2} x_2^{\ell} = s \tag{4.17}$$

$$y_{ih}^{\ell} \in \{0, 1\}, h \in H, i \in \{1, 2\}, \tag{4.18}$$

where L_i is the set of potential locations for firm i. p_{ih}^{ℓ} is the marginal delivered cost of firm i from location ℓ to market h. x_i is a vector with components x_i^{ℓ} with 0–1 values, where $x_i^{\ell} = 1$ indicates that facility location ℓ is chosen by firm i. For any fixed location set x_1 and x_2, y_{ih}^{ℓ} is a 0–1 variable indicating whether firm i serves market h from facility located at ℓ or not, and r and s is the number of facilities located by firm 1 and 2, respectively. Once the above problem is solved, the profit of each firm is obtained by taking into account that profit of a firm plus social cost equals total delivered cost of its rival [56]. A genetic like algorithm is proposed in [55] for the solution of the problem.

It is pointed out by [27, 30] that the socially optimal location does not necessarily guarantee the existence of an equilibrium in the first stage of the game, if demand is elastic with respect to price or if the marginal cost of production is not constant.

In such a case an equilibrium is obtained by an alternate procedure where each firm determines its optimal location considering the location of its rivals as given. Alternatively, the location-price problem is treated in [26] as a location game of an entering firm where the reaction of its rivals is to change their price instead of their locations. In other words, the entering firm first selects its locations and sequentially it gets involved in a Bertrand game with the existing firms aiming at maximizing its profit. Under reasonable assumptions the authors in [26] show that there is a price equilibrium to the Bertrand game in which each firm sets the optimal price in its monopolized markets and sets a price equal to the minimum delivered cost in other markets, i.e.,

$$p_h^*(\ell) = \begin{Bmatrix} p_h^{\text{mon}}(c_{\ell h}) & \textit{if} \;\; p_h^{\text{mon}}(c_{\ell h}) \leq c_h^{\text{com}} \\ c_h^{\text{com}} & \textit{if} \;\; p_h^{\text{mon}}(c_{\ell h}) > c_h^{\text{com}} \end{Bmatrix} \tag{4.19}$$

Consequently, if all competing firms set the equilibrium prices, the location-price problem reduces to the following location problem

$$\max \sum_{h \in H^*} \sum_{\ell \in L_h} q_h[p_h^*(\ell)][p_h^*(\ell) - c_{\ell h}]y_{\ell h} \tag{4.20}$$

$$s.t \sum_{\ell \in L_h} y_{\ell h} \leq 1, h \in H^* \tag{4.21}$$

$$y_{\ell h} \leq x_\ell, h \in H^*, \ell \in L_h \tag{4.22}$$

$$\sum_{\ell \in L} x_\ell = s \tag{4.23}$$

$$x_\ell, y_{\ell h} \in \{0.1\}, \tag{4.24}$$

where L is the set of potential location of the entering firm. L_h is the set of the location candidates at which the entering firm can price below its competitors, i.e. $L_h = \{\ell \in L : c_{\ell h} < c_h^{\text{com}}\}$. H^* is the set of markets in which the entering firm makes a positive profit, $H^* = \{h \in H : L_h \neq \emptyset\}$ x_ℓ and $y_{\ell h}$ are location and allocation variables, respectively, with $x_\ell = 1$ if a facility is located at ℓ; 0 otherwise, and $y_{\ell h} = 1$ if market h is served from location ℓ; 0 otherwise.

4.3 Location-Quantity Equilibria

The studies belonging to this case describe a non-cooperative game where competing firms decide on location and quantity supplied to market with goal to maximize their own profit. Because the choice of location is usually prior to decision, the location-quantity game is, in most of the cases, formulated as a two-stage game where in the first stage firms simultaneously decide their location and in the second stage give the chosen sets of location they decide the quantity to be offered at the market. Because of firms compete in quantities rather in price the second stage is a Cournot game.

Labbé and Hakimi [47] analyze a duopolistic game with linear demand and zero fixed cost in a network $N = (V, E)$ connecting spatially separated markets. A market is located at each vertex $v_k \in V$, and at each market a product is sold at price

$$p_k = \begin{cases} a_k - \beta_k q_k & \text{if } 0 \leq q_k \leq a_k/\beta_k \\ 0 & \text{otherwise} \end{cases}$$

with $a_k \geq 0$ and $\beta_k \geq 0$. Each firm locates at its facility points x_1 and x_2, respectively, and transport the product to the market v_k with unit transportation cost $T(d(x_i, v_k))$ which is assumed to be increasing with distance. The marginal cost of production $C(x_i)$ is assumed to be concave and independent from the quantity produced. Consequently, $c_k(x_i) = C(x_i) + T(d(x_i, v_k))$ is the unit delivered cost of firm i at market v_k. Labbé and Hakimi [47] solve the game using backward induction. First, the second stage of the game is solved for a fixed pair of location. They proved that there is a unique production pair,

$$q_{1k}^* = \arg\max_{q_{1k} \geq 0} \{(a_k - \beta_k(q_{1k} + q_{2k}^*) - c_k(x_i))q_{1k}\} \tag{4.25}$$

and

$$q_{2k}^* = \arg\max_{q_{2k} \geq 0} \{(a_k - \beta_k(q_{1k}^* + q_{2k}) - c_k(x_i))q_{2k}\} \tag{4.26}$$

which constitutes a Cournot-Nash equilibrium of the second stage of the game. Next, given the equilibrium quantities, the sub-game perfect Nash equilibrium is obtained by determining the equilibrium locations, i.e., by solving the first stage of the game. They state that if the unit delivered cost is sufficiently small so that is always profitable to offer some quantity at each market then there always exists a pair of equilibrium locations that consists of a pair of vertices for the first stage of the game. Sarkar et al. [68] extended the work in [47] by considering the case of oligopoly. They find similar condition for the existences of an equilibrium.

The game was further extended by Rhim et al. [62] by including capacity decisions as well. Their settings is a three-stage location game in which identical firms, first, decide their location, second, their capacity, and finally the production quantity for each market. Rhim et al. [62] demonstrated the existence of a sub-game Nash equilibrium by showing that the stage one game can be reduced into a single stage decision related to the so-called congestion games [65]. At equilibrium, the opened facilities select the markets they serve based on the demand, and the costs incurred by capacity acquisition, production, and transportation. The assumption of identical firm was relaxed by Sáiz and Hendrix [66], who assumed that the n

competitors have their own cost structure. Sáiz and Hendrix [66] presented the necessary optimality conditions for the Nash equilibria and developed algorithms to find them.

Konur and Geunes [41] adopt the common two-stage approach to study the effect of traffic congestion on competitive firms' equilibrium facility location and supply quantity decision. They study a set of k competitive firms considering the location of facilities at m possible locations in order to serve customer markets at n locations. Each firm incurs transportation, traffic congestion, and fixed facility location costs as a result of their location and distribution volume decisions. The transportation cost is linear in the quantity flow from facility $i \in I = \{1, 2, \ldots, m\}$ to market $j \in J = \{1, 2, \ldots, n\}$, with unit transportation cost c_{ij}.

The congestion cost is a non-decreasing convex function of the total quantity flow on the link (i, j) of the form $g_{ij}(q_{ij\bullet}) = \alpha_{ij} q_{ij\bullet}$, where $q_{ij\bullet}$ is the total quantity shipped from location i to market j, $q_{ij\bullet} = \sum_{r \in R} q_{ijr}$ and q_{ijr} is the quantity shipped from the facility of firm $r \in R = \{1, 2, \ldots, k\}$ at location i to market j. α_{ij} denotes the traffic congestion cost factor.

Konur and Geunes [41] conclude that firms will choose identical facility locations in equilibrium and that increased congestion hampers efficient location of facilities since it forces firms to use either congested links or links that are not close to the market. These results are further extended to the case of no-identical firms in [42].

Chapter 5
Competitive Facility Location with Competition of Customers

The research work dealing with the competitive location problems is limited only to the competition among the locators, that is, it is supposed that either both the locator and the allocator are the same or the customer knows the optimality criterion of the locator and agrees passively with it. Customers preferences as well as externalities (such as road congestion, facility congestion, emissions, etc.) caused by the location decisions are either ignored or "controlled" by incorporating constraints in order to "ensure" the achievement of a predetermined target. However, this approach treats customers as irresolute beings. Thus, if, for example, the customers travel to the facilities to obtain the offered service, then there is no compulsion or incentive for them to attend the designated facility. This means that, once the facilities are open, what the locator wishes the customers to do may not coincide with their own wish and behavior.

5.1 Market Competition and Location Decisions

The first attempt to study the influence of market competition on location decisions is done by Tobin and Friesz [75]. They analyze the case of a profit maximizing firm which is entering into spatially separated markets and knows that its location decisions will have impact on market prices.

It is assumed that the firm of interest wishes to establish production facilities at a set of eligible nodes N_0 so as to maximize its profits. The firm's profit at a node $\ell \in N_0$ is

$$Z_Q = \pi_\ell Q_\ell - V_\ell(Q_\ell) - F_\ell \tag{5.1}$$

A. Karakitsiou, *Modeling Discrete Competitive Facility Location*,
SpringerBriefs in Optimization, DOI 10.1007/978-3-319-21341-5_5

where π_ℓ is the market price at ℓ, Q_ℓ is the production level of the facility located at ℓ, $V_\ell(Q_\ell)$is the total variable cost of production at ℓ, and F_ℓ is the fixed cost of locating at ℓ. The firm's location problem can be stated as

$$\max Z(y, Q, \pi) = \sum_{\ell \in N_0} [\pi_\ell Q_\ell - V_\ell(Q_\ell) - y_\ell F_\ell] \tag{5.2}$$

$$st\ Q_\ell \leq \bar{Q}_\ell y_\ell, \forall \ell \in N_0 \tag{5.3}$$

$$\sum_{\ell \in N_0} Q_\ell \leq \bar{Q} \tag{5.4}$$

$$Q_\ell \leq 0, \forall \ell \in N_0 \tag{5.5}$$

$$y_\ell \in \{0, 1\}, \forall \ell \in N_0 \tag{5.6}$$

$$\pi = \phi(Q). \tag{5.7}$$

Constraints (5.3) require that the production level at ℓ to be less than the capacity $\bar{Q}_\ell$ if the facility is located at $\ell(y_\ell = 1)$ or zero if it is not $(y_\ell = 0)$. Constraint (5.4) imposes a limitation on the total level of production, constraints (5.5) require the production variables to be non-negative, and constraints (5.6) require choice variable to be zero or one. Finally, it is required that the market is in an equilibrium given production levels $Q_\ell, \forall \ell \in N_0$ and the equilibrium price is π_ℓ. The implicit relationship between market price and production level Q is given by (5.7).

One of the major difficulties of problem (5.2)–(5.7) is that the constraint (5.7) is not known explicitly. The only way to determine a price vector π given a production vector Q is to solve the spatial competition model.

To address the problem they proposed two different models to capture the market competition and its effect on price and production quantities: a spatial price equilibrium (SPE) which determines equilibria in price and production levels for perfectly competitive market and a Cournot Nash oligopolistic model in which a few profit maximizing firms compete in spatially separated markets. They used sensitivity analysis on variational inequalities to relate changes in location to changes in production to obtain optimal locations.

This model was generalized in [23] to allow the locating firm to determine shipping patterns in addition to production quantities and location. Due to mathematical complexity of these problems Miller et al. [52] proposed a heuristic algorithm based on sensitivity analysis methods which presume the existence of a solution and which locally approximate price changes as linear functions of production perturbations resulting from newly established facilities are presented.

The model proposed by Nagurney et al. [53] combines both the concepts of Nash equilibrium and SPE in order to develop the supply chain network equilibrium (SCNE) model, for investigating the economic behaviors of the players in a decentralized supply chain with the market competition. Although the SCNE model does not consider location decision, it is able to find the equilibrium shipment and price patterns between manufacturers and retailers, which would become necessary inputs in modeling the competitive facility location problem. Indeed, Meng et al. [51] have successfully developed the SCNE with production capacity, which is formulated by variational inequalities and employ the logarithmic-quantratic proximal prediction-correction (LQP P-C) method, developed by He et al. [31] as a solution algorithm. Next, the authors, building upon this model, proposed a generic mathematical program with equilibrium constraints for the competitive facility location, which can simultaneously determine facility location of the entering firm and the production levels of these facilities so as to optimize an objective.

To construct the model it is supposed that there are m manufacturers, n retailers, and o demand markets for an existing decentralized supply chain. All manufacturers produce a kind of product and supply them to the demand markets via the retailers. L be the number of candidate sites where the entering firm's facilities can be built. Moreover, it is assumed that the new built facilities are treated as additional manufacturers joining the existing decentralized supply chain, and consequently, competition arises among these new manufacturers as well as the existing manufacturers. It is further assumed that the entering firm is able to predict SCNE shipment and price patterns between the manufacturers and the retailers after the new facilities joining the decentralized supply chain. The decision variables of the model are

- x_l a binary decision variable to represent whether or not location l will be selected by the entering firm, namely
- q_{ij} shipment of the product from manufacturer i to retailer j
- q_{jk} shipment of the product from retailer j to demand market k
- Q^1 a row vector of all shipments between manufacturers and retailers
- Q^2 a row vector of all shipments between retailers and demand markets, i
- ρ_{3k} unit price of the product that consumers at demand market k are willing to pay

and the following function and assumptions are made:

- $f_i(Q^1)$ production cost of manufacturer i, a convex and continuously differentiable function of all the shipments between manufacturers and retailers
- $c_{ij}(q_{ij})$ convex and continuously differentiable transaction cost function between manufacturer i and retailer j
- $c_{jk}(Q^2)$ strictly monotone increasing transaction cost function between retailer j and demand market k
- $c_j(Q^1)$ handling cost of retailer j, a convex and continuous function of all the shipments between retailers and demand markets
- $d_k(q_{3k})$ elastic demand function of the product at demand market k

If x is a row vector of all the binary decision variables $\tilde{Q}^1(x)$ and $\widetilde{\rho_i}(x)$ are the row vectors of all SCNE shipments and prices between manufacturers, $i \in M(x)$ and retailers $j = 1, \ldots, n$, where $M(x)$ is the set of all manufacturers associated with the decision variable x,

$$M(x) = \{l | x_i = 1, l = 1 \ldots, L\} \cup \{L + 1, \ldots, L + m\} \tag{5.8}$$

then the problem is formulated as

$$\max_x h(x, \tilde{Q}^1(x), \widetilde{\rho_1}(x)) \tag{5.9}$$

st

$$q_p(x, \tilde{Q}^1(x), \tilde{\rho_1}(x) \leq 0, p = 1, \ldots, P, \tag{5.10}$$

$$\sum_{i \in M(x)} \sum_{j=1}^{n} \left[\frac{\partial f_i(\tilde{Q}^1(x))}{\partial q_{ij}} + \frac{\partial c_{ij}(\tilde{q}_{ij}(x))}{\partial q_{ij}} + \frac{\partial c_j(\tilde{Q}^1(x))}{\partial q_{ij}} \right] \times \left[q_{ij} - \tilde{q}_{ij} \right]$$

$$+ \sum_{j=1}^{n} \sum_{k=1}^{o} \left[c_{jk}(Q^2(x)) - \rho_{3k}(x) \right] \times \left[q_{jk} - q_{jk}(x) \right] +$$

$$+ \sum_{k=1}^{o} \left[\sum_{j=1}^{n} q_{jk}(x) - d_k(\rho_3(x)) \right] \times \left[\rho_{3k} - \rho_{3k}(x) \right] \leq 0$$

$$\forall (Q^1, Q^2, \rho_3) \in \Omega(x) \times R^0_+ \tag{5.11}$$

$$x_l = \{0, 1\}, l = 1, \ldots, L \tag{5.12}$$

where $\Omega(x)$ is the set of all shipment patterns for the supply chain involving the new built facilities. This set can be expressed as

$$\Omega(x) = \left\{ (Q^1, Q^2) \in \Re^{m_x n + no} \middle| \; \sum_{j=1}^{n} q_{ij} \leq C_i, i \in M(x), \quad \sum_{k=1}^{o} q_{jk} \leq \sum_{i \in M(x)} q_{ij}, j = 1, \ldots, n \right\}, \tag{5.13}$$

where m_x is the cardinality of the set $M(x)$, and the analytical expression of $\tilde{\rho}_1(x)$, which is a row vector of the SCNE unit price charge by a manufacturer for a retailer $\tilde{\rho}_{1ij}(x)$, is

$$\tilde{\rho}_{1ij}(x) = \frac{\partial f_i(\tilde{Q}^1(x))}{\partial q_{ij}} + \frac{\partial c_{ij}(\tilde{q}_{ij}(x))}{\partial q_{ij}} + \lambda_i(x), i \in M(x), j = 1, \ldots, n, \tag{5.14}$$

with $\lambda_i(x)$ be the optimal Lagrangian multiplier of the parameterized Variational Inequality (5.11).

The objective function (5.9) is a generic objective function and it can be customized according to the different requirements in the competitive facility location problem. Constraints (5.10) are also generic expression of different considered constraints in the competitive facility location problem, such as resource constraint. Equation (5.11) is the variational inequality which describes the SCNE conditions.

A hybrid genetic algorithm GA that incorporates with the LQP P-C is developed by Meng et al. [51] for solving problem (5.9)–(5.12). In the proposed algorithm chromosomes are encoded by the row vector of the binary variables x. The existence of the constraints in (5.10) leads to the computational difficulty for GA to check the feasibility of a chromosome, to tackle the problem the penalty function approach (see [4]) is employed. Constraints (5.10) are transformed into the objective function by using a penalty term ω and the fitness function is given by

$$H(x) = h(x, \tilde{Q}^1(x), \tilde{\rho}_1(x)) - \omega \sum_{p=1}^{P} \max\{0, q_p(x, \tilde{Q}^1, \tilde{\rho}_1(x))\}, \tag{5.15}$$

(5.11) is solved by LQP P-C algorithm.

In [35] a bi-level programming model is presented to seek the optimal location for logistics distribution centers. The upper-level model is to determine the optimal location by minimizing the planners' cost and the lower gives an equilibrium demand distribution by minimizing the customers' cost

$$\min \sum_{i=1}^{m} \sum_{j=1}^{n} C_{ij}(X_{ij})X_{ij} + \sum_{j=1}^{n} f_j z_j \tag{5.16}$$

$$st \quad \sum_{J=1}^{n} z_j \geq 1 \tag{5.17}$$

$$z_j \in \{0, 1\} \tag{5.18}$$

where X_{ij} solves

$$\min \sum_{i=1}^{m} \sum_{j=1}^{n} \int_0^{X_{ij}} D^{-1}(w)dw \tag{5.19}$$

$$st \quad \sum_{j=1}^{n} X_{ij} = w_i, \forall i = 1, \ldots, m, \tag{5.20}$$

$$\sum_{i=1}^{m} X_{ij} \leq s_j, \forall j = 1, \ldots, n, \tag{5.21}$$

$$X_{ij} \leq M z_j, \forall i = 1, \ldots, m, j = 1, \ldots, n, \tag{5.22}$$

$$X_{ij} \geq 0, \forall i = 1, \ldots, m, j = 1, \ldots, n, \tag{5.23}$$

where $C_{ij}(\cdot)$ is the unit generalized cost of meeting the demand of customer i from the distribution center j, and it is usually a nonlinear function; X_{ij} is the demand of the customer i supplied by distribution center j; f_j is the fixed investment associated with building distribution center j; z_j is a $0-1$ variable, if distribution center j is built, then z_j takes the value of 1, and 0 otherwise; $D^1(\cdot)$ is the inverse of demand functions; w_i is the total demand of customer i; s_j is the capacity of distribution center j; M is an arbitrarily large positive constant.

From the point of decision-makers, the first term on the objective function (5.16) represents the total costs of meeting customers demand. Constraint (5.17) ensures that at least one distribution center is built, and constraint (5.18) represents the binary restrictions of the decision variables. The lower-level problem represents the customers' choice behaviors. Constraint (5.20) ensures that the total demand of each customer must be met by supplying from some distribution centers. Constraints (5.21) are the capacity constraints, which ensure that all the demand distributed in a distribution center will not exceed its capacity. Constraint (5.22) prohibits the demand on any proposed distribution center that is not actually constructed. Based on the special form of constraints (5.22), a simple reaction function is proposed. This reaction function is obtained by transforming (5.22) into the form

$$X_{ij} = Mz_j - y_{ij}^* \tag{5.24}$$

where y_{ij}^* is the optimal relaxation variable obtained after solving the second-level problem by any existing algorithm. This reaction function is substituted in the first level of the problem which results to an integer programming problem with variables $\mathbf{z}$ which can be solved by any well-known nonlinear programming model.

5.2 Costumer Competition and Location Decisions

In [38–40] the effects of customers' competition for the offered service level on the facility location decisions are examined. Two types of decision-makers are considered, the producer who tries to provide at facilities the best level of service at minimum cost and the customers who make their choices in order to minimize their perceived costs. The customers are involved in a Nash-type game in their effort to ensure the best level of services for themselves. A bi-level programming model is formulated in order to take into consideration the effects of customers' competition. Furthermore an extension is also proposed. It is assumed that there are two producers who constitute a duopoly in the network. The producers compete with each other with respect to the service level they offer in order to attract customers. A bi-level model with two leaders is proposed in order to take into account both the competition between producers and the competition among customers.

The evaluation of the offered service is based on the delay faced by the customers at each distribution center i. If x_{ij} is the amount that the customer j buys from the distribution center i, then the performance function $d_i(x_i)$ measures the level of service offered by the distribution center i where $x_i = \sum_{j=1}^{n} x_{ij}$. Suppose that m is the set of potential sites for the location of the distribution centers. We assume that the establishment of a distribution center to the candidate site i implies a fixed location cost f_i. Furthermore, suppose r_j is the demand of customer $j(j = 1 \ldots, n)$, p_i is the unit price paid by customers, and q_1 is the capacity of the distribution center $i(i = 1 \ldots, m)$. Under the assumption that a central coordinator chooses the location of the distribution center in such a manner that the total cost of the system is minimized, the mathematical model can be formulated as follows:

$$\textbf{(SO} - \textbf{FL)} \ \min \quad \sum_{i=1}^{m} d_i(x_i, y_i) + \sum_{j=1}^{m} p_i x_i + \sum_{i=1}^{n} \sum_{j=1}^{m} t_{ij} x_{ij} + \tag{5.25}$$

$$+ \sum_{i=1}^{m} F_i y_i$$

$$\text{s.t} \quad \sum_{i=1}^{m} x_{ij} = r_j, \ \forall j \tag{5.26}$$

$$x_i \leq y_i q_i, \ \forall i \tag{5.27}$$

$$x_i - \sum_{j=1}^{n} x_{ij} = 0, \ \forall i \tag{5.28}$$

$$y_i \in \{0, 1\}, \forall i \tag{5.29}$$

$$x_{ij} \geq 0, \ \forall i, \ \forall j \tag{5.30}$$

The objective function of problem (5.26) minimizes the total cost consisting of the cost of the delay, plus the transportation and purchasing costs plus the cost involved in setting up a distribution center. Constraints (5.26) ensure that the quantities purchased by the customer j at all distribution centers meet his overall demand. Constraints (5.27) impose that the total amount of the product available at each distribution center i does not exceed its capacity. In addition, it enables that the assignment of the customers' demand only in sited distribution. Relations (5.28) are the defining constraints of the model, ensuring the maintenance of flow in the network.

In a second model the producer takes into account the free will and the competitive preference of the customers and determines the final location of the distribution centers based on the prediction of their behavior as delivered by the outcome of a Nash game. Thus, the problem is formulated as bi-level programming model:

$$\text{(BSO} - \text{FL)} \min_{[y_i]} \quad \sum_{i=1}^{m} F_i y_i + \sum_{i=1}^{m} d_i(\bar{x}_i, y_i)\bar{x}_i + $$

$$+ \sum_{i=1}^{m} p_i \bar{x}_i + \sum_{i=1}^{m}\sum_{j=1}^{n} t_{ij}\bar{x}_{ij} \tag{5.31}$$

$$\text{s.t} \quad y_i \in \{0, 1\}, \ \forall i \tag{5.32}$$

where $[\bar{x}_i]$ and $[\bar{x}_{ij}]$ solve

$$\text{(UO} - \text{TP)} \min \quad \sum_{i=1}^{m} \int_0^{x_i} d_i(t)dt$$

$$+ \sum_{i=1}^{m} p_i x_i + \sum_{i=1}^{m}\sum_{j=1}^{n} t_{ij} x_{ij} \tag{5.33}$$

$$\text{s.t} \quad \sum_{i=1}^{m} x_{ij} = r_j, \ \forall j \tag{5.34}$$

$$x_i \leq q_i y_i, \ \forall i \tag{5.35}$$

$$x_i - \sum_{j=1}^{n} x_{ij} = 0, \ \forall i \tag{5.36}$$

$$x_{ij} \geq 0 \ \forall i, j \tag{5.37}$$

According to this model, the leader (producer) decides the location of distribution centers solving problem (5.31) and (5.32), but he does not control the variables x_i and x_{ij} since they describe the choices of his customers. The values of the variables $[\bar{x}_i]$ and $[\bar{x}_{ij}]$ are derived from the model (5.33)–(5.37) corresponding to an oracle. In other words, the leader uses (5.33)–(5.37) as an oracle to discover trends/reactions of the customers in each potential location and tries to minimize the total cost of the system based on these discoveries.

In cases where there are more than one producers, none of them have the power to direct customers to distribution centers. Thus, as a result, the offered service level and the customer satisfaction are the basic differentiation and discrimination components among economic units of the same sector. In order to take into account both levels of competition we formulate the following bi-level problem with two leaders.

Let us assume that the potential location of distribution centers $i = 1, \dots, m$ is dispersed between the two producers who in turn are involved in a competition for customer attraction through the provided service level. Let M_1 and M_2 ($m = |M_1| + |M_2|$) be the nodes of the two producers, respectively. Then, under the assumption that both producers "announce their strategies simultaneously," we obtain a Nash game with two players who are dealing (for $K = 1, 2$) with the following problems:

The facility location problem of the producer 1:

$$(\mathbf{CFL_1})\ \min \quad \sum_{i\in M_1} F_i y_i +$$

$$+\sum_{i\in M_1} d_i(\bar{x}_i, y_i)\bar{x}_i + \sum_{i\in M_1} p_i\bar{x}_i + \sum_{i\in M_1}\sum_{j=1}^{n} t_{ij}\bar{x}_{ij} \tag{5.38}$$

$$\text{s.t} \quad y_i \in \{0,1\}, \forall i \in M_1 \tag{5.39}$$

The facility location problem of the producer 2:

$$(\mathbf{CFL_2})\ \min \quad \sum_{i\in M_2} F_i y_i +$$

$$+\sum_{i\in M_2} d_i(\bar{x}_i, y_i)\bar{x}_i + \sum_{i\in M_2} p_i\bar{x}_i + \sum_{i\in M_2}\sum_{j=1}^{n} t_{ij}\bar{x}_{ij} \tag{5.40}$$

$$\text{s.t} \quad y_i \in \{0,1\}, \forall i \in M_2 \tag{5.41}$$

where $[\bar{x}_i]$ and $[\bar{x}_{ij}]$ solve (5.33)–(5.37) .

The producers compete with each other with respect to the service level they offer in order to attract customers involved in a Nash game. A Nash equilibrium for this duopolistic game corresponds to a set of location and capacity choices (strategies), which ensure that none of the players are better off by unilaterally changing his strategy.

Let $Y = \{y | y_i \in \{0,1\}, \forall i \in M_k\}$, be the feasible sets of the players for $k = 1, 2$, $\mathbf{y}_k = [y_i]_{i\in M_k}$ and $\mathbf{y} = \begin{bmatrix}\mathbf{y}_1\\ \mathbf{y}_2\end{bmatrix}$. We have already mentioned the existence of optimal solutions $\bar{x}_i$ and $\bar{x}_{ij}$ for given capacity $[\bar{q}_i]$. Thus, there is a function from $\mathbb{R}^m$ to $\mathbb{R}^m$, such that for a given $\bar{\mathbf{y}}$ it returns the unique equilibrium point $[\bar{x}_i]$ from (5.33)–(5.37) and a corresponding mapping from $\mathbb{R}^m$ to $\mathbb{R}^{m\cdot n}$ such that for a given $\bar{\mathbf{y}}$ it returns an optimal transportation plan $[\bar{x}_{ij}]$ which corresponds to the equilibrium point $[\bar{x}_i]$, thus it holds that $\bar{x}_i = x_i(\bar{\mathbf{y}})$ and $\bar{x}_{ij} = x_{ij}(\bar{\mathbf{y}})$, respectively.

Hence problems $(\mathbf{CFL_k})$ could be formulated as a single level problems:

$$(\mathbf{SCFL_k})\ \min_{\mathbf{y}_k\in Y_k} \quad \sum_{i\in M_k} d_i(x_i(\mathbf{y}), y_i)x_i(\mathbf{y}) + \sum_{i\in M_k} p_i x_i(\mathbf{y}) + \tag{5.42}$$

$$+\sum_{i\in M_k}\sum_{j=1}^{n} t_{ij}x_{ij}(\mathbf{y}) \tag{5.43}$$

Each problem, $(\mathbf{SCFL_k})$ corresponds to player k who is involved into the Nash game.

References

1. R. Aboolian, O. Berman, D. Krass, Competitive facility location model with concave demand. Eur. J. Oper. Res. **181**, 598–619 (2007)
2. R. Aboolian, O. Berman, D. Krass, Competitive facility location and design problem. Eur. J. Oper. Res. **182**, 40–62 (2007)
3. E. Alekseeva, N. Kocheva, Y. Kotcetov, A. Plyasunov, Heuristic and exact methods for the discrete $(r|p)$-centroid problem, in *Evolutionary Computation in Combinatorial Optimization*, ed. by P. Cowling, P. Merz. Lecture Notes in Computer Science, vol. 6022 (Springer, Berlin, 2010), pp. 11–22
4. M.S. Bazaraa, H.D. Sherali, C.M. Shetty, *Nonlinear Programming: Theory and Algorithms* (Wiley, New York, 1993)
5. V.L. Beresnev, Upper bounds for objective function of discrete competitive facility location problems. J. Appl. Ind. Math. **3**(4), 3–24 (2009)
6. V. Beresnev, Branch and bound algorithm for a competitive facility location problem. Comput. Oper. Res. **40**, 2062–2070 (2013)
7. V.L. Beresnev, On the competive facility location problem with free choice of supplier. Autom. Remote Control **75**(4), 668–676 (2014)
8. V.L. Beresnev, A.A. Mel'nikov, Approcimate algorithms for the competitive facility location problem. J. Appl. Ind. Math. **5**(2), 180–190 (2011)
9. O. Berman, D. Krass, Locating multiple competitive facilities: spatial interaction models with variable expenditures. Ann. Oper. Res. **111**, 197–225 (2002)
10. C.M. Campos Rodríguez, D.R. Santos Peñate, J.A. Moreno Pérez, An exact procedure and lp formulations for the leader-follower location problem. TOP **18**(1), 97–121 (2010)
11. R. Church, C. ReVelle, The maximal covering location problem. Pap. Reg. Sci. **32**(1), 101–118 (1974)
12. S. Consoli, J.A. Moreno-Pérez, N. Mladenovic, K. Darby-Dowman, Discrete article swarm optimization for the minimum labeling spanning tree problem. Nat. Comput. **9**, 29–46 (2010)
13. I.A. Davydov, Y.A. Kochetov, N. Mladenovic, D. Urosevic, Fast metaheuristics for the discrete $(r|p)$-centroid problem. Autom. Remote Control **75**(4), 677–687 (2014)
14. G. Dobson, U.S. Karmarkar, Competitive location on a network. Oper. Res. **35**(4), 565–574 (1987)
15. P. Dorta-González, D.R. Santos-Pe nate, R. Suárez-Vega, Spatial competition in networks under delivered pricing. Pap. Reg. Sci. **84**(2), 271–280 (2005)
16. T. Drezner, Locating a single new facility among existing facilities unequally attractive facilities. J. Reg. Sci. **34**, 237–252 (1994)

A. Karakitsiou, *Modeling Discrete Competitive Facility Location*,
SpringerBriefs in Optimization, DOI 10.1007/978-3-319-21341-5

17. T. Drezner, Competitive facility location in plane, in *Facility Location. A Survey of Applications and Methods*, ed. by Z. Drezner (Springer, Berlin, 1995), pp. 285–300
18. T. Drezner, Competitive facility location, in *Encyclopedia of Optimization*, ed. by C. Floudas, P.M. Pardalos (Springer, New York, 2001), pp. 270–275
19. H.W. Drezner, Z. Hamacher (eds.), *Facility Location Applications and Theory* (Springer, Berlin, 2002)
20. H. Eiselt, G. Laport, J.F. Thisse, Competitive location models: a framework and bibliography. Transp. Sci. **27**, 44–54 (1993)
21. J. Fernández, S. Salhi, G.T. Boglárka, Location equilibria for a continuous competitive facility location problem under delivered pricing. Comput. Oper. Res. **41**, 185–195 (2014)
22. K. Fischer, Sequential discrete *p*-facility models for competitive location planning. Ann. Oper. Res. **111**, 253–270 (2002)
23. T.L. Friesz, R.L. Tobin, T. Miller, Existance theory for spatially competitive network facility location models. Ann. Oper. Res. **18**, 267–276 (1989)
24. T. Friez, T. Miller, R. Tobin, Competitive network facility location models: a survey. Pap. Reg. Sci. Assoc. **65**, 47–57 (1988)
25. P.M.D. García, P. Fernández, B. Pelegrín, On price competition in location-price models with spatially separated markets. TOP **12**(2), 351–374 (2004)
26. A.D. García, B. Pelegrín, P. Fernántez, Location strategy for a firm under competitive delivered prices. Ann. Reg. Sci. **47**, 1–23 (2011)
27. B. Gupta, Competitive spatial price discrimination with strictly convex production costs. Reg. Sci. Urban Econ. **24**(2), 265–272 (1994)
28. S.L. Hakimi, On locating new facilities in a competitive enviroment. Eur. J. Oper. Res. **12**, 29–35 (1983)
29. S.L. Hakimi, Location with spatial interaction, in *Discrete Location Theory*, ed. by P.B. Mirchandani, R.L. Francis (Wiley, New York, 1990), pp. 439–478
30. J.H. Hamilton, Spatial discrimination: bertrand vs. cournot in a model of location choice. Reg. Sci. Urban Econ. **19**(1), 87–102 (1989)
31. B.S. He, Y. Hu, X.M. Yuan, A logarithmic-quadratic proximal prediction-correction method for structured monotone variational inequalities. Comput. Optim. Appl. **35**, 19–46 (2006)
32. M.J. Hodgson, A location-allocation model maximizing consumers welfare. Reg. Stud. **15**, 493–506 (1981)
33. H. Hotelling, Stability in competition. Econ. J. **39**, 41–57 (1929)
34. D. Huff, Defining and estimating a trade are. J. Mark. **28**, 34–38 (1948)
35. S. Huijun, G. Ziyou, W. Jianjun, A bi-level programming model and solution algorithm for the location of logistics distribution centers. Appl. Math. Model. **32**, 610–616 (2008)
36. M.G. Ierapetritou, Single facility location: multi-objective rectilinear distance location, in *Encyclopedia of Optimization*, ed. by C. Floudas, P.M. Pardalos (Springer, New York, 1999), pp. 3613–3617
37. M.G. Ierapetritou, Single facility location: multi-objective euclidean distance location, in *Encyclopedia of Optimization*, ed. by C. Floudas, P.M. Pardalos (Springer, New York, 2009), pp. 3610–3613
38. A. Karakitsiou, Coordination and competition in supply chain: optimization and game theoretic approaches. Ph.D. thesis, Technical University of Crete, 2007 (in Greek)
39. A. Karakitsiou, Competitive multilevel capacity allocation, in *Optimization Theory, Decision Making, and Operations Research Applications*, ed. by A. Migdalas et al. Springer Proceedings in Mathematics & Statistics (Springer, Berlin, 2013)
40. A. Karakitsiou, A. Migdalas, Locating facilities in a competitive environment. Optim. Lett.
41. D. Konur, J. Geunes, Analysis of traffic congestion costs in a competitive supply chain. Transp. Res. E: Logist. Transp. Rev. **47**(1), 1–17 (2011)
42. D. Konur, J. Geunes, Competitive multi-facility location games with non-identical firms and convex traffic congestion costs. Transp. Res. E **48**(1), 373–385 (2012)
43. D. Kress, E. Pesch, Sequential competitive location on networks. Eur. J. Oper. Res. **217**(3), 483–499 (2012)

44. H. Küçkayadin, N. Aras, I.K. Altinel, A discrete competitive facility location model with variable attractiveness. J. Oper. Res. Soc. **62**, 1726–1741 (2011)
45. H. Küçkayadin, N. Aras, I.K. Altinel, Competitive facility location problem with attractiveness adjustment of the follower: a bilevel programming model and its solution. Eur. J. Oper. Res. **208**, 206–220 (2011)
46. H. Küçkayadin, N. Aras, I.K. Altinel, A leader-follower game in competitve facility location. Comput. Oper. Res. **39**, 437–448 (2012)
47. M. Labbé, S.L. Hakimi, Market and locatinal equilibrium for two competitors. Oper. Res. **80**(1), 749–756 (1991)
48. P. Lederer, J.F. Thisse, Competitive location on networks under delivere pricing. Oper. Res. Lett. **9**, 147–153 (1990)
49. A.A. Mel'nikov, Computational complexity of the discrete competititive facility location. J. Appl. Ind. Math. **8**(4), 557–567 (2014)
50. A.A. Mel'nikov, Randomized local search for the discrete competitive facility location problem. Autom. Remote Control **75**(4), 700–714 (2014)
51. Q. Meng, Y. Huang, R.L. Cheu, Competitive facility location on decentralized supply chain. Eur. J. Oper. Res. **196**, 487–499 (2009)
52. T. Miller, Heuristic algorithms for delivered price spatially competitive network facility location problem. Ann. Oper. Res. **34**, 177–202 (1992)
53. A. Nagurney, J. Dong, D. Zhang, A supply chain network equilibrium model. Transp. Res. E **38**, 281–303 (2002)
54. M. Nakanishi, L.G. Cooper, Parameters estimation for a multiplicative competitive interaction model: least squares approach. J. Mark. Res. **11**, 303–311 (1974)
55. B. Pelegrín, J.L. Redondo, P. Fernández, I. García, P.M. Ortigosa, Gasub: finding global optima to discrete location problems by a genetic-like algorithm. J. Global Optim. **38**(2), 249–264 (2007)
56. B. Pelegrín-Pelegrín, P. Dorta-González, P. Fernández-Hernádez, Finding location equilibria for competing firms under delivered pricing. J. Oper. Res. Soc. **62**, 729–741 (2011)
57. F. Plastria, Static competitive facility location: an overview of optimisation approaches. Eur. J. Oper. Res. **129**(3), 461–470 (2001)
58. F. Plastria, L. Vanhaverbeke, Discrete models for competitive location with foresight. Comput. Oper. Res. **35**(3), 683–700 (2008)
59. J.L. Redondo, J. Fernández, I. García, P.M. Ortigosa, Parallel algorithms for continuous multifacility competitive location problems. J. Global Optim. **50**(4), 557–573 (2011)
60. C. ReVelle, The maximum capture or sphere of influence problem: hotelling revised on a network. J. Reg. Sci. **26**, 343–357 (1986)
61. C. ReVelle, D. Serra, The maximun capture problem including reallocation. Inf. Oper. Res. **29**, 130–138 (1991)
62. H. Rhim, T.H. Ho, U.S. Karmarkar, Competitive location, production, and market selection. Eur. J. Oper. Res. **149**(1), 211–228 (2003)
63. M.C. Roboreto, A.A. Pessoa, A branch and cut algorithm for the discrete $(r|p)$-centroid problem. Eur. J. Oper. Res. **224**, 101–109 (2013)
64. C.C. Rodríguez, J.A. Moreno-Pérez, D.R. Santos-Pe nate, Particle swarm optimization with two swarms for the discrete $(r|p)$-centroid problem, in *Computer Aided Systems Theory – EUROCAST 2011*, ed. by R. Moreno-Díaz, F. Pichker, A. Quesada-Arencibia. Lecture Notes in Computer Science (Springer, Berlin, 2011), pp. 432–439
65. R. Rosenthal, A class of games possesing pure-strategy nash equilibria. J. Game Theory **2**(1), 65–67 (1973)
66. M.E. Sáiz, E.M.T. Hendrix, Methods for computing nash equilibria of a location-quantity game. Comput. Oper. Res. **35**(10), 3311–3330 (2008)
67. D.R. Santos-Pe nate, R. Suárez-Vega, P. Dorta-González, The leader-follower location model. Netw. Spat. Econ. **2007**, 45–61 (2007)
68. J. Sarkar, B. Gupta, D. Pal, Location equilibrium for cournot oligopoly in spatially seperated markets. J. Reg. Sci. **37**(2), 195–212 (1997)

69. D. Serra, C. ReVelle, Market capture by two competitors: the preemptive location problem. J. Reg. Sci. **34**, 549–561 (1994)
70. D. Serra, C. ReVelle, Competitive location in discrete space, in *Facility Location. A Survey of Applications and Methods*, ed. by Z. Drezner (Spribger, Berlin, 1995), pp. 367–386
71. D. Serra, C. ReVelle, Competitive location and pricing on networks. Geogr. Anal. **31**, 109–129 (1999)
72. J. Spoerhase, H.-C. Wirth, (r, p)-centroid problems on path and trees. Theor. Comput. Sci. **410**, 5128–5137 (2009)
73. R. Suárez-Vega, D.R. Santos-Peñate, P. Dorta-González, Competitive multi-facility location on networks: the $(r|x_p)$-medianoid problem. J. Reg. Sci. **44**(3), 569–588 (2004)
74. R. Suárez-Vega, D.R. Santos-Pe nate, P. Dorta-González, Discreatization and resolution of the $r|x_p$ medianoid problem involving quality criteria. TOP **12**(1), 111–133 (2004)
75. R. Tobin, T.L. Friesz, Spatial competition facility location models: definitions, formulations and solution appproach. Ann. Oper. Res. **6**, 49–74 (1986)
76. H. von Stackelberg, *The Theory of the Market Economy* [Translated from the German and with an Introduction by Alan T. Peacock] (William Hodge, London, 1952)
77. G. Wesolowsky, J. Brimberg, Optimizing facility location with rectilinear distancesoptimizing facility location with rectilinear distances, in *Encyclopedia of Optimization*, ed. by C. Floudas, P.M. Pardalos (Springer, New York, 2001), pp. 1903–1905

MIX
Papier aus verantwortungsvollen Quellen
Paper from responsible sources
FSC® C105338
FSC
www.fsc.org

If you have any concerns about our products,
you can contact us on
ProductSafety@springernature.com

In case Publisher is established outside the EU,
the EU authorized representative is:
Springer Nature Customer Service Center GmbH
Europaplatz 3, 69115 Heidelberg, Germany

Printed by Libri Plureos GmbH
in Hamburg, Germany